Iron Porphyrins, Part II

List of Contributors

Karl M. Kadish, Department of Chemistry, University of Houston, Houston, Texas (pp. 161–249)

Samaresh Mitra, Chemical Physics Group, Tata Institute of Fundamental Research, Homi Bhabha Road, Colaba, Bombay, India (pp. 1–42)

Graham Palmer, Department of Biochemistry, Rice University, Houston, Texas (pp. 43–88)

Thomas G. Spiro, Department of Chemistry, Princeton University, Princeton, New Jersey (pp. 89–159)

PHYSICAL BIONORGANIC CHEMISTRY SERIES

A. B. P. Lever and Harry B. Gray, Series Editors

NUMBER

1. A.B.P. Lever and Harry B. Gray (eds.), *Iron Porphyrins, Part I*, 1982

2. A.B.P. Lever and Harry B. Gray (eds.), *Iron Porphyrins, Part II*, 1982

Other numbers in preparation

Iron Porphyrins, Part II

Edited by

A.B.P. LEVER
York University
Downsview, Ontario, Canada

HARRY B. GRAY
California Institute of Technology
Pasadena, California

1983

ADDISON-WESLEY PUBLISHING COMPANY
Advanced Book Program/World Science Division
Reading, Massachusetts

London　·　Amsterdam　·　Don Mills, Ontario　·　Sydney　·　Tokyo

Coden: PBCSD3

Library of Congress Cataloging in Publication Data (Revised)
Main entry under title:

Iron porphyrins.

(Physical bioinorganic chemistry series; 1–2)
Includes bibliographies and indexes.

1. Heme. 2. Porphyrin and porphyrin compounds.
3. Oxidation-reduction. I. Lever, A. B. P. (Alfred Beverley Philip) II. Gray, Harry B. III. Series.
QP671.H45I76   574.19′2   82-6786
ISBN 0-201-05816-2 (v. 1)   AACR2
ISBN 0-201-05817-0 (v. 2)

Copyright © 1983 by Addison-Wesley Publishing Company, Inc.
Published simultaneously in Canada.

All rights reserved. No part of this publication may be reproduced, stored in a retrieval system, or transmitted, in any form or by any means, electronic, mechanical photocopying, recording, or otherwise, without the prior written permission of the publisher, Addison-Wesley Publishing Company, Inc., Advanced Book Program/World Science Division, Reading, Massachusetts 01867, U.S.A.

Manufactured in the United States of America

ABCDEFGHIJ-MA-898765432

CONTENTS

Series Editors' Foreword xi
Preface for Parts I and II xiii

1.
MAGNETIC SUSCEPTIBILITY OF IRON PORPHYRINS 1
Samaresh Mitra

A. Introduction 1

B. Iron(III) Porphyrins 4
    i. Theoretical Foundation 4
    ii. High-Spin Iron(III) Porphyrins 9
    iii. Low-Spin Iron(III) Porphyrins 24
    iv. Spin-Mixed Iron(III) Porphyrins 25
    v. Spin Equilibrium 29

C. Iron(II) Porphyrins 31
    i. Theoretical Foundations: The Ligand-Field Model 32
    ii. Examples 35

D. Concluding Remarks 39

Acknowledgments 39

References 39

2.
ELECTRON PARAMAGNETIC RESONANCE OF HEMOPROTEINS ... 43
Graham Palmer

A. Introduction 45
    i. Relevant EPR Concepts 45

B. EPR of Low-Spin Ferri-hemoproteins 52
    i. The Interpretation of Low-Spin Ferric Heme EPR Spectra 55
    ii. Highly Anisotropic Low-Spin Hemoproteins 59
    iii. Which Way Is "y"? 61
    iv. Consequences of Field-Swept Spectra 63

C. EPR of High-Spin Hemoproteins 64

D. Mixed-Spin States 70

E. The EPR of Compound I of Peroxidase 73

F. EPR of Nitrosyl Hemoproteins 77

Acknowledgments 84

References 85

3. THE RESONANCE RAMAN SPECTROSCOPY OF METALLOPORPHYRINS AND HEME PROTEINS 89
Thomas G. Spiro

Abbreviations 91

A. Introduction 91

B. Resonance Enhancement 92
   i. Basic Concepts 92
   ii. Porphyrin π-π^* Transitions 94
   iii. Out-of-Plane Enhancement 102

C. Vibrational Assignments 114
   i. Porphyrin Skeletal Modes 114
   ii. Peripheral Substituent Effects 121
   iii. Axial-Ligand Modes 128

D. Structure Correlations 133
   i. Core-Size Correlations 134
   ii. π-Backbonding 138
   iii. Oxidation-State Marker 140

E. Protein Effects 142
   i. Metal-Ligand Modes 142
   ii. Porphyrin Skeletal Modes 145

F. Time-Resolved Studies 147

Acknowledgments 151

References 152

4. THE ELECTROCHEMISTRY OF IRON PORPHYRINS IN NONAQUEOUS MEDIA 161
Karl M. Kadish

A. Introduction 161

B. Techniques 166
   i. Polarography 166
   ii. Cyclic Voltammetry 169

Contents ix

    iii. Coulometry and Controlled-Potential Electrolysis 175
    iv. Joint Application of Electrochemical and ESR
        Techniques 177
    v. Spectroelectrochemistry 180

C. **Solvent Systems and Supporting Electrolytes** 181
    i. Aqueous Solutions 181
    ii. Aprotic Solvents 182
    iii. Supporting Electrolytes 184

D. **Solvent and Counterion Effect on Reversible
Half-Wave Potentials** 186
    i. Fe(III) ⇌ Fe(II) 186
    ii. Fe(II) ⇌ Fe(I) 191
    iii. Oxidation of Fe(III) 193

E. **Axial Ligation, Redox Potentials, and Calculation of
Formation Constants** 194
    i. Iron(II) Complexation and the Reaction Fe(II) ⇌ Fe(I) 195
    ii. Iron(III) Complexation and the Reaction Fe(III) ⇌ Fe(II) 199
    iii. Mixed Axial Ligation of Fe(III) 207
    iv. Diatomic-Molecule Adducts 208
    v. Oxidation of Fe(III) 209

F. **Porphyrin Structure and Half-Wave Potentials** 209
    i. Substituent Effects on Half-Wave Potentials 211
    ii. Measurements of Electron-Transfer Rates 219

G. **Oxidation-Reduction Mechanisms** 221
    i. Reduction Mechanism in DMF 221
    ii. Reduction Mechanism in Me_2SO 225
    iii. Reduction Mechanism in CH_2Cl_2 and $EtCl_2$ 228

H. **Variable-Temperature Electrochemistry** 229

I. **Electrode Reaction of Iron Dimers** 232
    i. Oxidation of Oxo-Bridged Dimers 232
    ii. Reduction of Oxo-Bridged Dimers 233
    iii. Nitrido-Bridged Dimer 234
    iv. Axial-Ligand-Binding Reactions 235

J. **Comparisons Between Iron Complexes and Other
Metalloporphyrins** 236
    i. Effect of Solvent on $E_{1/2}$ for M(III) ⇌ M(II) 237
    ii. Effects of Ligand Complexation on $E_{1/2}$ for
       M(III) ⇌ M(II) 239
    iii. Effects of Counterion on $E_{1/2}$ for M(III) ⇌ M(II) 240
    iv. Heterogeneous Electron-Transfer Rate Constants 241

K. Characterization of Reactants and Products 241

Acknowledgments 244

References 244

Index 251

Series Editors' Foreword

In recent years there has been explosive growth in the study of metal ions in living systems, a field commonly referred to as "bioinorganic chemistry". This field crosses the boundaries of inorganic, organic, and physical chemistry, chemical physics, biophysics, biochemistry, and medicine. It demands of its practitioners a broadly based knowledge covering many of these areas. The field is vast but there have been few textbooks other than some introductory texts and some early monographs. Most "bioinorganickers" were trained in inorganic chemistry or in biochemistry. Surprisingly few have come from physics and related areas. It is the goal of this Series to provide monographs that will stimulate the entrance of more chemical physicists into the field. To this end a somewhat pedagogical tone will be struck, and the emphasis of the first few volumes will be on the use of physical techniques in the study of oxidation-reduction proteins.

Oxidation-reduction reactions are pervasive features of metalloprotein chemistry. Such processes occur ubiquitously in all aspects of life from bacteria to plants to mammals. Important examples include electron transfer in cytochromes and in iron-sulfur proteins, reduction of molecular oxygen to water, incorporation of oxygen atoms in substrates, and incorporation of nitrogen into growing plants.

These first books will show how theory allied with modern physical techniques can be used to unravel the mysteries and intricacies of biological oxidation-reduction. The early volumes deal exclusively with biomolecules containing iron, molybdenum, and copper, which currently appear implicated in more than 90% of the known biological redox processes.

The common theme of the series of volumes is: "What can we learn about the geometric structures, the electronic structures, and the mechanisms of biological redox centers through the use of modern physical techniques?"

Sufficient background information on the chemical physics of simple iron, copper, and molybdenum systems is provided to form a basis for discussion of the more complex protein environment.

Successive volumes are planned to deal with heme proteins, iron-sulfur and other non-heme iron redox proteins, copper proteins, and molybdenum proteins. Two further volumes are currently planned. One will deal primarily with the mechanisms of biological redox reactions and the other is planned to be a text for a graduate course in physical bioinorganic chemistry. The latter will summarize in one volume all the important aspects of bioinorganic redox chemistry encountered in detail in earlier volumes.

The editors have emphasized the need to include only hard factual information and accepted theory and avoid inclusion of speculative materials so that this series will remain current and valid for many years.

The Series is open-ended. Readers are invited to contact the Editors with suggestions for further volumes.

A.B.P. LEVER
HARRY B. GRAY

Preface for Parts I and II

1964 saw the publication of a book destined to become a classic in the field of metalloporphyrins. This book "Porphyrins and Metalloporphyrins", written by J. E. Falk, covered much of the information known at the time. Partly spurred by this book, the field expanded rapidly and the need arose for a new edition of 'Falk'. Regretfully, Falk passed away and the second edition was not completed. Instead, the need was met by Kevin Smith, who, in 1976 edited a new edition of "Porphyrins and Metalloporphyrins". Almost 900 pages long, this book highlighted many of the developments since 1964 and emphasized synthesis and physical techniques.

Bioinorganic chemistry has continued to develop rapidly and porphyrin chemistry has grown with it. This series of volumes, "Physical Bioinorganic Chemistry", dedicated to consideration of the redox active proteins, is produced to attempt to keep pace with this development and to spur further development. Within the porphyrin field itself, the iron porphyrins, ubiquitous in the animal kingdom, fulfill roles in electron transfer proteins, oxygen transport proteins and in catalysts for oxygen insertion into biological substrates. They clearly play a central role. These two volumes are devoted specifically to the iron porphyrins, and specifically to an understanding of their electronic and geometric structure, leading, it is to be hoped, to better recognition of the factors determining these various roles. In size these volumes, devoted only to iron, approach that of Smith, devoted to the entire porphyrin field.

Nevertheless, despite this rapid progress, the books are written with the intent that they will remain current and useful for many years. The presentation is mildly pedagogical, to show how physical techniques can be used to understand the electronic and geometric structure of the iron porphyrins. All the iron porphyrins known in the biological arena are covered, studies supplemented by recourse also to the synthetic iron porphyrins. However, to maintain utility for many years, speculation is minimized and the various Chapters emphasize information known to be true and hence lasting.

The Chapter sequence in these two volumes arises partly as a consequence of pedagogy and partly as a consequence of availability of manuscripts at the publication deadline for Volume I, which will appear one month before Volume II. Volume I begins with a theoretical and structural basis (Loew, then Scheidt/Gouterman) thus laying a foundation for the techniques which follow. The clearest justification for the theoretical molecular orbital studies, discussed by Loew, is probably presented by the electronic spectra of these species, presented by Makinen and Churg in Chapter 3. The nature of the iron porphyrin bond is nicely studied through nuclear magnetic resonance spectroscopy, as discussed by Goff in Chapter 4, completing this volume.

Volume II opens with a review of the static magnetic properties of these species (Mitra), properties which can also be readily verified with theory. This Chapter (II.1) naturally leads into electron spin resonance spectroscopy (Palmer) II.2), a powerful tool in this area. Resonance Raman spectroscopy, a comparatively new technique, has significant value in this field, linking iron electronic states with porphyrin structure (Spiro, Chapter II.3). This is followed by Electrochemistry (Kadish, II.5), providing insight into the tuning of iron states by variations in porphyrin structure, and providing clues into the redox activity of these species.

It is to be expected that the insight gained by the use and development of these various techniques will facilitate our understanding of the complex roles played by iron porphyrins in living systems.

These volumes are written at a level assuming undergraduate knowledge of inorganic and physical chemistry. They should prove of value to inorganic, organic and physical chemists, biochemists, and medical workers from graduate level to established researchers. Since the iron porphyrins play such a vital role in living systems, these books should prove of value to anyone interested in the field, from the inorganic theoretician to the bio-technologist.

A.B.P. LEVER
H. B. GRAY

Iron Porphyrins, Part II

1
Magnetic Susceptibility of Iron Porphyrins

SAMARESH MITRA

A. INTRODUCTION

Over the years there has been a continuing and ever growing interest in the physical and chemical properties of iron porphyrins. This has resulted in the accumulation of enormous amount of information on this important class of molecules. Extensive review articles exist on X-ray crystallography, ESR, NMR, Mössbauer, and optical spectroscopy, and several other aspects of iron porphyrins [1-7]. Surprisingly, no detailed review articles have yet been written on the magnetic susceptibility of iron porphyrin or heme proteins. The magnetic susceptibility is a very fundamental measurement and is immensely informative, especially if the measurements are done on single crystals and over a wide range of temperature [8, 9]. Information obtained from magnetic-susceptibility measurements often complement or supplement those obtained from other physical techniques. The purpose of this article is to assess critically such information on iron porphyrins and examine the advantages and limitations of magnetic-susceptibility studies for this purpose. To achieve this objective we have selected a few representative examples of each spin situation for our discussion; no attempt has been made to make this article an up-to-date reference listing.

The measurement of magnetic susceptibility is relatively straightforward, and the methods are well documented in the literature. The average magnetic susceptibility ($\bar{\chi}$) can be conveniently measured by Faraday, Foner, or SQUID magnetometers which require small amount of sample. The SQUID magnetometer is particularly suitable for measurements on proteins, where very large sensitivity is needed. However a note of caution has to be given with regard to measurement of the average magnetization, especially at high magnetic fields and low temperatures. Under these conditions the polycrystallites have a tendency to orient preferentially in the magnetic field due to the inherent large anisotropy of the system, thereby vitiating the experimental results. To overcome this problem the polycrystallites should be dispersed in a diamagnetic vaseline mull and cooled down to low temperatures so that the crystallites remain randomly oriented.

Single-crystal susceptibilities of iron porphyrins can be conveniently determined by measuring their magnetic anisotropy with the help of a torque magnetometer. The principle and details of the technique are described elsewhere [8, 10]. Generally, single crystals weighing between 2 and 10 mg are required, depending on the magnitude of anisotropy. Measurement yields directly the difference between the crystal susceptibilities which, when combined with average susceptibility, gives the principal crystal susceptibilities. Magnetic-anisotropy measurement is however by itself very useful, as we shall see later. The crystal anisotropy must be transformed into the molecular anisotropy, which is the quantity of interest to chemists. This transformation is achieved through tensor algebra as described in the literature [10, 11] and is quite simple in axial symmetry. For most metalloporphyrins the symmetry of the crystal field around the metal ion is nearly axial even though the overall symmetry of the molecule may be much lower. In our discussions on iron porphyrins we shall assume axial symmetry unless otherwise warranted. A necessary precaution in the analysis of the experimental anisotropy data is to recognize the importance of a large diamagnetic anisotropy due to the delocalized porphyrin macro-ring, which must be considered for obtaining the true paramagnetic anisotropy. This is generally achieved by correcting the experimental data, either by using the anisotropy of an isostructrual (or at least analogous) diamagnetic porphyrin or by computing the diamagnetic anisotropy of such a system.

We shall denote the principal molecular susceptibility and moment by K_i and μ_i ($i = \parallel$ or $\perp$) respectively. The subscript $\parallel$ or $\perp$ refers to the quantities parallel or perpendicular to the symmetry axis of the molecule. Thus by our definition $K_\parallel$ (or $\mu_\parallel$) will be perpendicular to, and $K_\perp$ (or $\mu_\perp$) will be parallel to the plane of the porphyrin molecule. The experimental anisotropies will often be plotted as $\mu_\perp^2 - \mu_\parallel^2$, which is related to the principal susceptibility through $\mu_i^2 = 7.995 K_i T$. The magnetization σ is generally expressed as reduced moment, $\langle \mu \rangle = \sigma/N\beta$, in conformity with the Brillouin function [12]. Throughout this chapter the susceptibility will be expressed in units of 10^6 cm^3/mole and the magnetic moment in Bohr magnetons.

B. IRON(III) PORPHYRINS

Iron(III) porphyrins have been the subject of the most extensive studies, as they often show magnetic properties remarkably similar to ferric heme proteins [13]. Most of the ferric porphyrins can be fairly easily synthesized and can be crystallized in stable and pure form. They exhibit a variety of spin states as well as spin-mixed ground states. Some iron(III) porphyrins show behavior characteristic of thermal spin equilibrium. All these situations resemble closely the properties of iron(III) heme proteins, and have made the iron(III) porphyrins a good model for studying and understanding several properties of biomolecules [5, 6].

We shall discuss first the theoretical foundations for rationalizing the magnetic properties (with emphasis on magnetic susceptibility) of iron(III) porphyrins. We shall use the crystal-field (or ligand-field) model, which has provided the necessary theoretical basis for this purpose. Some selected examples of each spin state will then be discussed separately, as well as those showing "unusual" spin behavior.

i. Theoretical Foundations

A ferric ion with d^5 electronic configuration can exist in a ground state of high ($S = \frac{5}{2}$), low ($S = \frac{1}{2}$), or intermediate ($S = \frac{3}{2}$) spin. There is also the possibility that the ground state of the ferric ion may not be a pure spin state but may contain substantially admixed components of more than one spin state. In some special circumstances the ferric ion may be in the spin "crossover" region and show thermal spin equilibrium. We discuss below, on the basis of a crystal-field model essentially due to Harris [13–15], the factors which govern these possibilities, and we outline the magnetic properties of the ferric porphyrins with a variety of ground-state configurations.

A d^5 ferric ion possesses a total of 43 high-, low-, and intermediate-spin multiplet states in a crystal field of cubic symmetry. Harris has shown, however, that out of these only the three lowest-lying strong-field terms— $^2T_{2g}(t_{2g}^5)$, $^4T_1(t_{2g}^3 e_g^2)$, and $^6A_1(t_{2g}^3 e_g^2)$—mix significantly via spin-orbit coupling and hence need consideration. In a crystal field of strongly tetragonal symmetry, as in ferric porphyrins, the splitting and energies of the resulting states in terms of the Racah parameters B and C and crystal-field parameters Δ_{Oh}, C', and μ' defined by Harris are shown in Figure 1. Δ_{Oh} is the cubic crystal-field parameter and arises from the averaged effect of all six ligands of a given complex. Both C' and μ' relate to differences between the axial and in-plane field strengths, defined as positive when the in-plane interaction exceeds the axial and negative when the axial interaction is greater. For iron porphyrins C' and μ' are generally expected to be positive.

For an iron(III) ion we can now assign regions of parameter space to various ground states. Using free-ion values for B, C, and ζ (the one-electron spin-orbit coupling constant), Figure 2 defines the regions of various possible spin ground states together with the region of spin-mixed ground state. As is seen from the figure, the high spin state can be stabilized for both positive and negative values of C', but negative values are not likely for reasons stated above. A pure quartet ground state can be stabilized if $\Delta_{Oh} \geq 36,000$ cm^{-1} for $C' \geq 9,000$ cm^{-1}, indicating a very weak axial interaction compared to the in-plane one. The low spin state can be achieved for all values of Δ_{Oh} in the strong-axial-field region. Here again positive values of C' are more likely, and hence ^2E is the physically more meaningful ground state. For values of

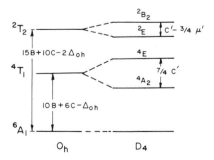

Figure 1 The first-order electrostatic and crystal-field interactions for the sextet, quartet, and doublet states of the d^5 configuration in O_h and D_4 symmetry.

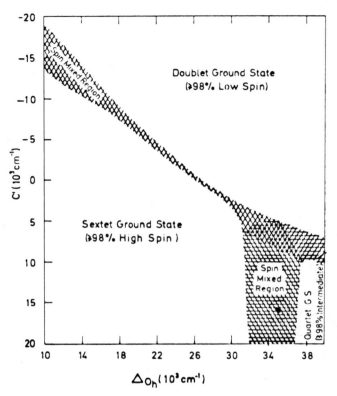

Figure 2 Parameter-space mapping defining regions of ground states with pure and mixed spin states for a d^5 ferric ion. Calculation was done taking free-ion values of B, C, and ζ ($= 1100$, 3750, and 420 cm^{-1} respectively) [13].

$\Delta_{Oh} \leqslant 31{,}000$ cm^{-1}, spin mixing of ground and low-lying states occurs as the axial field is decreased, but the region of the spin-mixed wave functions is very narrow. However, this region is appreciably large for $36{,}000 \geqslant \Delta_{Oh} \geqslant 31{,}000$ cm^{-1} and C' large and positive.

In Figure 3–5 we show typical dependences of the average and principal magnetic moments ($\bar{\mu}$, $\mu_\parallel$, $\mu_\perp$) on Δ_{Oh} and C' at 300 and 4 K, spanning the various regions of parameter space. One point immediately becomes clear: that variations in both Δ_{Oh} and C' can generate $\bar{\mu}$ varying from 5.9 (high spin) to $\simeq 2.0$ BM (low spin) and cover the intermediate values, which have been experimentally observed for ferric heme proteins. Let us now discuss separately the expected magnetic behavior in these regions.

In the high-spin regions two observations are clear. First, $\mu_\perp > \mu_\parallel$ in the positive region of C', and the anisotropies are large for higher values of Δ_{Oh}. A change in the sign of C' changes the sign of the anisotropy. Since $\mu_\perp$ is generally larger than $\mu_\parallel$, C' (as expected) is positive for iron(III) porphyrins.

Iron(III) Porphyrins

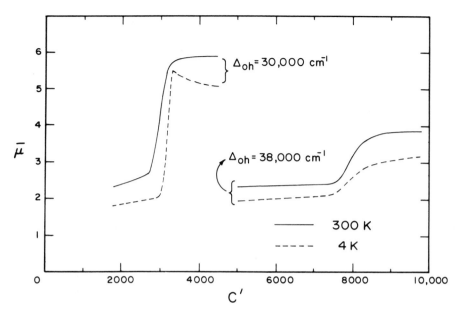

Figure 3 Dependence of average magnetic moment on C' at 300 and 4 K for two typical values of Δ_{Oh}.

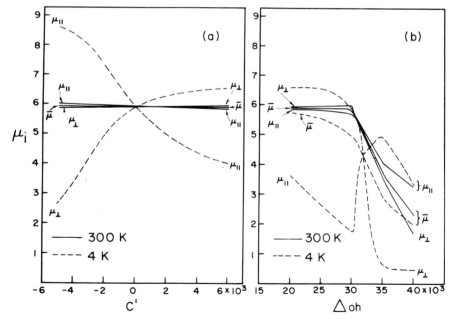

Figure 4 Dependence of the principal and average magnetic moments at 300 and 4 K on C' and Δ_{Oh}: (a) $\Delta_{Oh} = 20,000$ cm^{-1}, (b) $C' = 7,000$ cm^{-1}.

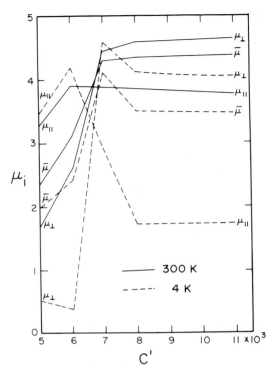

Figure 5 Dependence of the average and principal magnetic moments at 300 and 4 K on C' for $\Delta_{Oh} = 34{,}000$ cm^{-1}.

Second, the principal moments depend quite markedly on Δ_{Oh} and C' (except when C' is very large) but are rather insensitive to $\bar{\mu}$ (cf. Figures 3 and 4): hence the advantage of the single-crystal measurement.

In the high-spin region the ground state is $^6A_1$ and the excited states generally lie much higher in energy. Hence the magnetic properties can be described reasonably well by a spin Hamiltonian of the form

$$\mathcal{H} = DS_z^2 + g\beta \vec{H} \cdot \vec{S} \tag{1}$$

where D is the zero-field splitting (ZFS) parameter. The spin-Hamiltonian formalism has been quite extensively used to describe the magnetic properties of iron porphyrins, presumably for the sake of its simplicity [8, 16]. The zero-field splitting in Equation (1) is related to the crystal-field parameters above, which we shall discuss later.

In the pure low- and intermediate-spin regions the dependence of the average and principal magnetic moments is typical of these spin situations. However, the experimental results for ferric porphyrins with these spin states are too limited to warrant a detailed discussion. In the spin-mixed region the magnetic properties of the ferric ion depend very markedly on Δ_{Oh} and C' (cf. Figures 3 and 5), and the magnetic moments show anomalous variation with temperature. We shall discuss this aspect in some detail later.

Though Harris's model has been successful in the interpretation of the magnetic properties of heme systems, it suffers from a drawback. While the model has been applied to systems with strong tetragonal distortion, it uses octahedral (i.e. cubic) wave functions corresponding to $^6A_{1g}$, $^4T_{1g}$, and $^2T_{2g}$ for the calculation of matrix elements and hence the magnetic properties. It has recently been shown [17] that this assumption leads to serious error in the choice of the wave function for the ^4E term and gives inappropriate matrix elements of the spin-orbit coupling. A more generalized treatment of a d^5 ion in tetragonal symmetry spanning a wide range of tetragonal distortion has therefore been presented [17]. Fortunately however the above discrepancy in Harris's model does not appear to cause significant error for the iron heme systems, since the contribution of the ^4E term in these cases is generally negligible.

ii. High-Spin Iron(III) Porphyrins

The magnetic susceptibility and magnetization of a large number of synthetic and natural high-spin iron(III) porphyrins have been studied over a wide range of temperature and magnetic field. Measurements have been made on both single-crystal and polycrystalline samples, and the results provide a good test of the theory as well as a comparison with the analogous heme proteins. The magnetic properties of ferric porphyrins have been studied by several other methods. The magnetic susceptibility study therefore provides a good opportunity to ascertain the usefulness of the technique for heme systems in comparison with other physical methods.

a. Tetraphenylporphyrin Iron(III) Halides. The five-coordinate high-spin tetraphenylporphyrin iron(III) halides, Fe(TPP)X, start our discussion. These synthetic porphyrins can be easily obtained in pure form and crystallized as fairly large single crystals. X-ray structural data are available on several members of this series which establish close similarity in the Fe—N_4X chromophores (Figure 6 and Table 1). Extensive data on the average magnetic susceptibility [18–21], magnetic anisotropy [21–24], high-field magnetization [18–21], Mössbauer spectroscopy [25, 26], far-infrared spectroscopy [27, 28], electron spin resonance [29], and NMR [30–32] have been reported. A typical temperature dependence of the average magnetic moment [19, 20] is shown in Figure 7. The magnetic moment near room temperature is close to the spin-only value of 5.9 BM and remains nearly constant down to $\simeq$ 40 K. The sharp decrease in $\bar{\mu}$ at lower temperatures is characteristic of nearly all high-spin iron(III) porphyrins and is indicative of large zero-field splitting of

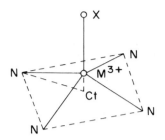

Figure 6 Fe—N_4X chromophore in the Fe(TPP)X system.

the ground state. Thus the average-susceptibility data at lower temperature may be used to deduce the ground-state zero-field splitting parameter [16], though it is not very sensitive to it and may lead to large errors (see below). The single crystals of the Fe(TPP)X (X = Cl, Br, I, NCS) show the effect of this large zero-field splitting even at room temperature [22–24]. The room- and low-temperature paramagnetic-anisotropy data summarized in Table 2 bring out the highly anisotropic character of these porphyrins and a trend in its variation with the axial ligand. Since the paramagnetic anisotropy in these systems arises almost wholly from the zero-field splitting [8], $g_\parallel$ and $g_\perp$ being very nearly equal to 2 here, the magnitude and trend in the variation of the paramagnetic anisotropy essentially bears a direct correspondence to the zero-field splitting. The temperature dependence of the paramagnetic anisotropy for the two typical members of the series is shown in Figures 8 and 9, which also indicate that the anisotropy arises almost wholly from the zero-field splitting.

The magnetization measurements at high magnetic fields and low temperatures are very sensitive to the ground-state properties of the iron(III) heme

Table 1. X-Ray Structural Data for Fe(TPP)X

	Fe—N (Å)	Fe—X (Å)	Fe—ct (Å)
Fe(TPP)NCS[a]	2.065	1.957	0.55
Fe(TPP)Cl[b]	2.060	2.192	0.39
Fe(TPP)Br[c]	2.069	2.348	0.56
Fe(TPP)I[d]	2.066	2.554	0.53

[a] Hoard, J. L., and Bloom, M., 173rd Am. Chem. Soc. Natl. Meeting, New Orleans, 1977, Abst. 27.
[b] Hoard, J. L., Cohen, G. H., and Gillick, M. D., *J. Am. Chem. Soc.* **89** (1967) 1992.
[c] Skelton, B. W., and White, A. H., *Aust. J. Chem.* **30** (1977) 2655.
[d] Hatano, K., and Scheidt, W. R., *Inorg. Chem.* **18** (1979) 877.

Iron(III) Porphyrins

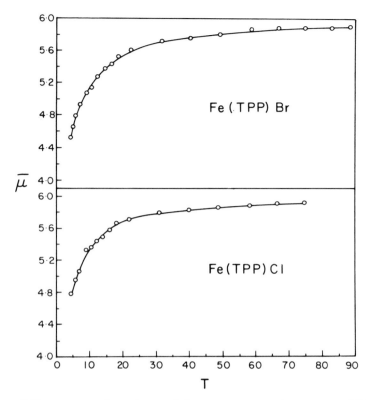

Figure 7 Temperature dependence of the average magnetic moment of Fe(TPP)X (X = Cl and Br). Data taken from References [19] and [20].

systems [9, 19–24]. Several such early measurements [18, 33] were, however, inaccurate and limited, and suffer from the errors mentioned in the Introduction. Recently very accurate and detailed magnetization measurements have been reported for Fe(TPP)Cl and Fe(TPP)Br in the range 2–20 K and 10–50 kOe [19, 20]. The magnetization deviates considerably from the Brillouin function for $S = \frac{5}{2}$ in both cases, indicating again the large zero-field splitting

Table 2. Paramagnetic Anisotropy ($K_\perp - K_\parallel$) of the Fe(TPP)X Series at Two Temperatures [a]

Temp. (K) \ Compound	Fe(TPP)NCS [24]	Fe(TPP)Cl [22]	Fe(TPP)Br [20]	Fe(TPP)I [24]
296.0	1,020	1,425	2,920	3,150
85.0	13,800	15,200	31,200	32,100

[a] A diamagnetic correction of $(K_\perp - K_\parallel)_{dia} = 540 \times 10^{-6}$ was employed.

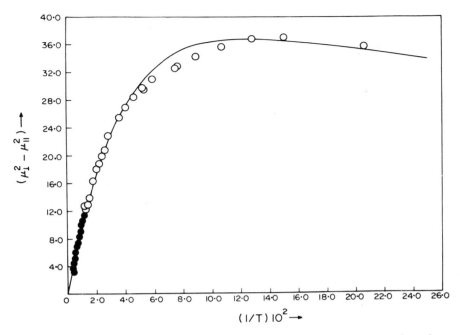

Figure 8 Temperature dependence of the molecular anisotropy $(\mu_\perp^2 - \mu_\parallel^2)$ of Fe(TPP)Cl between 295 and 5 K. The filled circles represent the data in the 295–80-K range. The data are taken from References [22] and [23]. The solid curve is the theoretical one. (See text.)

(cf. Figure 10). A detailed set of typical data is shown in Figure 11 for Fe(TPP)Br. The magnetization for H ≥ 40 kOe shows almost complete saturation below 4 K. The saturation moment is however much lower than the value $\langle \mu \rangle = 5$ BM expected for $S = \frac{5}{2}$ without any zero-field splitting [12].

A preliminary discussion of the above data on the basis of the spin-Hamiltonian formalism is quite instructive. Using Equation (1), we plot the variation of $\Delta\mu^2$ as a function of the ZFS parameter D at three typical temperatures (Figure 12). It is immediately clear that the sign of D is directly related to the sign of the paramagnetic anisotropy, i.e., whether $\mu_\perp^2 - \mu_\parallel^2$ is positive or negative. Further, the anisotropy, even in the liquid-nitrogen temperature range, is quite sensitive to D; at lower temperature it is far more so. Thus the paramagnetic anisotropy even in the 300–80 K temperature range is adequate to determine the sign and magnitude of D. The situation with respect to the average magnetic moment is not so clear. While it is recognized that the data at lower temperatures are sensitive (though to a lesser extent) to the magnitude of D, they may not by themselves uniquely determine its sign. Hence caution must be exercised in using the average magnetic-susceptibility

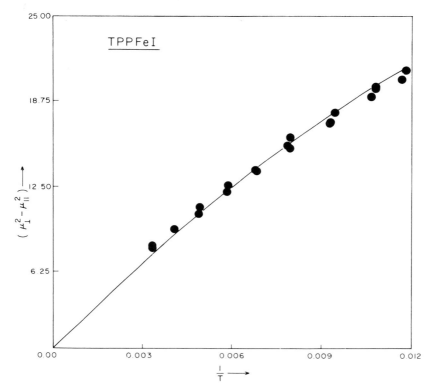

Figure 9 Temperature dependence of the molecular anisotropy of Fe(TPP)I. The solid line is the calculated one [24].

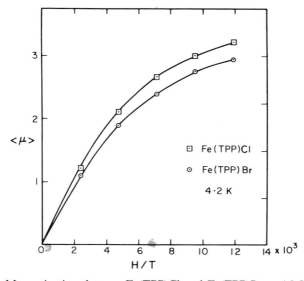

Figure 10 Magnetization data on Fe(TPP)Cl and Fe(TPP)Br at 4.2 K [19, 20].

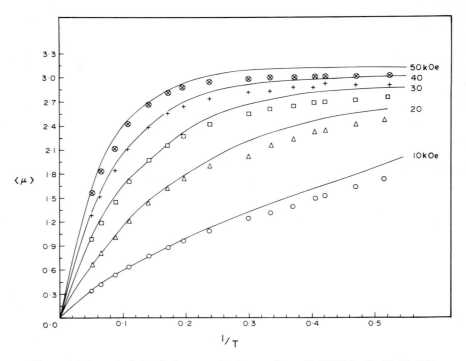

Figure 11 A typical detailed magnetization study on Fe(TPP)Br (2–20 K) [20].

data on high-spin ferric heme systems for such purposes. Using Van Vleck's equation for the magnetic susceptibility, explicit expressions for the principal and average magnetic moments can be deduced for Equation (1) as follows:

$$\mu_\parallel^2 = \frac{3g_\parallel^2}{4}\left\{\frac{1 + 9e^{-2x} + 25e^{-6x}}{1 + e^{-2x} + e^{-6x}}\right\}, \tag{2}$$

$$\mu_\perp^2 = \frac{3g_\perp^2}{4}\left\{\frac{9 + (16 - 11e^{-2x} - 5e^{-6x})/2x}{1 + e^{-2x} + e^{-6x}}\right\}, \tag{3}$$

where $x = D/kT$.

These expressions may only be used for analysis of the average and anisotropy data at high temperatures and low magnetic fields, though they have often been rather erroneously used for analyzing low-temperature susceptibility data obtained at high magnetic fields [18]. For analysis of the magnetization data as in Figures 10 and 11, Equations 2–3 cannot be employed, as they assume $\beta H/kT \ll 1$ and $\beta H/D \ll 1$, neither of which is valid for iron(III) porphyrins at high magnetic fields and low temperatures (i.e. in

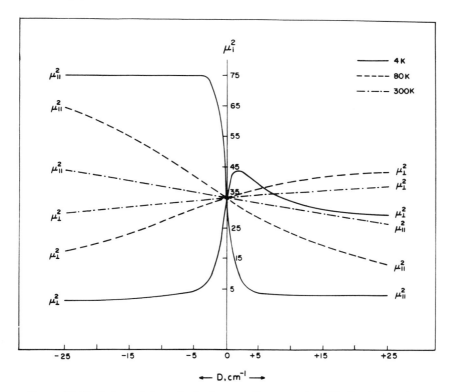

Figure 12 Variation of the principal magnetic moments with D at three temperatures [24].

the saturation region). The susceptibility in such cases must be calculated using the thermodynamic equation [34, 35]

$$\bar{\chi} = -\frac{N}{H} \frac{\Sigma(\partial E_i/\partial H)\exp(-E_i/kT)}{\Sigma \exp(-E_i/kT)}. \quad (4)$$

Care should also be exercised in calculating the average magnetization since the conventional method of averaging ($\bar{\sigma} = \frac{1}{3}\sigma_\parallel + \frac{2}{3}\sigma_\perp$) gives quite incorrect results in these limits [36]. Hence the average magnetization must be calculated using the spatial-averaging technique [36, 37].

Table 3 includes the values of D obtained by various methods for the Fe(TPP)X series. Fe(TPP)Cl has been studied by various methods, and the agreement in the D-values obtained from them is good except for Reference [18]. The slightly higher value obtained from the magnetization data is perhaps due to an antiferromagnetic exchange interaction which affects the magnetization at lower temperatures (see below). For Fe(TPP)Br, the value quoted in

Table 3. Zero-Field Splitting in Iron(III) Porphyrins and Heme Proteins

Compound[a]	D (cm^{-1})	Method	Ref.
Fe(TPP)Cl	6.0 ± 0.1	Magnetic anisotropy	[22]
	7.0 ± 1.0	Mössbauer spectroscopy	[26]
	8.0	Av. susceptibility and magnetization	[19]
	11.9	Av. susceptibility	[18]
	6.5	Far infrared	[27]
Fe(TPP)Br	12.5 ± 0.5	Magnetic anisotropy	[20]
	12.5	Av. susceptibility and magnetization	[20]
	4.9	Av. susceptibility	[18]
Fe(TPP)I	13.5	Magnetic anisotropy	[24]
Fe(TPP)NCS	5.0	Magnetic anisotropy	[24]
Fe(DPDME)X:			
X = Cl	8.95	Far infrared	[38]
X = Br	11.8	Far infrared	
X = I	16.4	Far infrared	
Hemin chloride	6.95	Far infrared	[38]
Fe^{3+} MbF	6.5	Magnetic anisotropy	[39]
Fe^{3+} metMb	10.5	Magnetic anisotropy	[39]

[a] DPDME, deuteroporphyrindimethylester; Mb myoglobin.

Reference [18] is again erroneous. The consistent errors there appear to be due to inaccuracies in the experimental data and its analysis as well as to the insensitive nature of the average susceptibility to D. A Mössbauer study in zero magnetic field is generally not useful for deducing quantitative information regarding D; measurement at high magnetic fields is required for this purpose. Far-infrared spectroscopy gives direct information about D. The single-crystal magnetic anisotropy, though not a direct method, is equally sensitive and accurate for this purpose. The high-field magnetization data can also be used for an accurate description of the ground-state properties, but they generally refer to very low temperatures and are often complicated by the presence of an antiferromagnetic exchange interaction.

The zero-field splitting in the Fe(TPP)X and related ferric porphyrins shows some systematic variation (Table 3). The variation in D across the halides for the same porphyrin series follows the sequence $D_{NCS} < D_{Cl} < D_{Br} \lesssim D_{I}$, which is the same as the spectrochemical series for these halides. The zero-field splitting also appears to significantly vary with substitutions on the porphyrin skeleton. For example, with chloride in the fifth apical position, the D-value varies from 6 cm^{-1} (TPP) to about 9 cm^{-1} (DPDME). Any trend in the variation in D with porphyrin substitution may be difficult to establish with the limited data available, and a more systematic study is required.

Let us now analyze the magnetic susceptibility data on the basis of the crystal-field model discussed in Section B and examine the information that can be derived from such measurements. A survey of the variation in the crystal-field parameters in the high-spin region shows that they are insensitive to $\bar{\mu}$ but sensitive to the single-crystal susceptibilities. Figure 4 shows that C' must be positive for the Fe(TPP)X series, since $\mu_\perp > \mu_\parallel$. A closer examination of the variation in μ_i in the various high-spin regions shows that Δ_{Oh} and C' ($= \mu'$) must lie close to 30,000 and 6,000 cm^{-1} respectively to reproduce the magnitude of the room-temperature anisotropy. The values of the crystal-field parameters for the Fe(TPP)X series deduced from the magnetic anisotropy and magnetization data are listed in Table 4. The differences between the two sets of values must be at least partly due to the reasons discussed above. It is interesting that the crystal-field parameters for these porphyrins lie very close to the "crossover" region in the parameter-space diagram (cf. Figure 2). Thus a small change in these parameters by a suitable chemical modification of the axial or equatorial ligand can force the ferric ion to move to low-spin or spin-mixed region. C' does not show the expected systematic variation with the halides, since in this region of parameter space magnetic data are rather insensitive to it. On the other hand, the magnetic data predict a systematic variation in Δ_{Oh}, suggesting that the change in the apical halogen affects the overall strength of the crystal field. It is interesting that the structural data do not however show any substantial change in the in-plane Fe—N bonding.

Using these values of the crystal-field parameters, the calculated eigenvalues and eigenfunctions for one of the representative member of the series are listed in Table 5. As expected, ^4E and the doublets lie much higher in energy and have very little effect on the magnetic properties of these porphyrins [40, 41]. It is the mixing with the low-lying $^4A_2$ which determines the magnetism. Further, the ground state $^6A_1$ remains almost pure $S = \frac{5}{2}$, the largest mixing in Fe(TPP)I being only about 4% of $^4A_2(\pm\frac{1}{2})$. The ground spin state is $^6A_1(\pm\frac{1}{2})$, consistent with the positive sign of D, and the splittings between the different

Table 4. Crystal-Field Parameters for the Fe(TPP)X Series [a]

X	From magnetization measurements		From anisotropy measurements	
	Δ_{Oh} (cm^{-1})	C' (cm^{-1})	Δ_{Oh} (cm^{-1})	C' (cm^{-1})
NCS	—	—	28,811	6,273
Cl	30,567	5,195	29,510	6,096
Br	31,724	6,115	31,519	6,300
I	—	—	31,587	6,202

[a] Reference [40].

Table 5. Eigenfunctions and Eigenvalues of Lowest-Lying States in Fe(TPP)I [a]

State	Main component	Energy (cm^{-1})	$^6A_1(\pm\frac{1}{2})$	$^6A_1(\pm\frac{3}{2})$	$^6A_1(\pm\frac{5}{2})$	$^4A_2(\pm\frac{3}{2})$	$^4A_2(\pm\frac{1}{2})$
Ground	$^6A_1(\pm\frac{1}{2})$	−118.1	0.9738	0.0000	0.0000	0.0000	0.2248
1 exc.	$^6A_1(\pm\frac{3}{2})$	−90.9	0.0000	0.9804	−0.0011	0.1917	0.0000
2 exc.	$^6A_1(\pm\frac{5}{2})$	−27.9	0.0000	0.0011	0.9989	0.0002	0.0000
3 exc.	$^4A_2(\pm\frac{3}{2})$	1887.8	0.0000	−0.1933	0.0002	0.9630	0.0000
4 exc.	$^4A_2(\pm\frac{1}{2})$	1989.4	−0.2250	0.0000	0.0000	0.0000	0.9694
5 exc.	$^2E(\mp\frac{1}{2},\pm 1)$	4592	—	—	—	—	—

[a] From Reference [40]. All other states lie much higher in energy.

low-lying spin states agree well with the magnitudes of the zero-field splitting deduced from the spin Hamiltonian. Table 5 also shows the reason why the spin Hamiltonian is a reasonably good approximation in the high-spin iron(III) porphyrins.

b. Hemin Chloride. Hemin chloride (protoporphynato iron(III) chloride) is structurally related to the Fe(TPP)X series [42] and has been the subject of several studies. An early measurement of the average susceptibility and magnetization down to 2 K and up to 20 kOe was interpreted on the basis of Equations (1)–(3), and a D-value of 11.5 cm^{-1} was deduced [18]. A subsequent measurement [43] of its average susceptibility down to 4 K could not however be fitted with Equations (2)–(3) for any value of D. The value of D in hemin chloride has been determined very accurately ($D = 6.95$ cm^{-1}) by far-infrared spectroscopy [38]. It has been pointed out [19, 33] that the above magnetization measurement suffers from the error mentioned in the Introduction and that the use of Equations (2)–(3) for the analysis of such data is incorrect [34]. Recently a very detailed and accurate measurement of the average susceptibility and magnetization [44] has been made in the range 2–20 K and 10–50 kOe, and a set of data is shown in Figure 13. The data fit Equation (1) very well when the susceptibility and magnetization are calculated using Equation (4) and the spatial-averaging technique. The zero-field splitting so deduced ($D = 8.0$ cm^{-1}) is however still slightly higher than the far-infrared value, though the agreement is much better than reported earlier. The small discrepancy in the D-value arises from the antiferromagnetic exchange interaction (see below).

c. Miscellaneous Examples. A detailed magnetization measurement at 4 K and up to 55 kOe has been reported for some model high-spin cytochrome P450 mono-oxygenase enzymes containing protoporphyrin IX as the prosthetic group [44a]. These iron(III) porphyrins are also five-coordinated with a thiolate or phenolate axial ligand. The thiolate is especially interesting because axial ligation through the sulphur is speculated to occur in some cytochromes. The

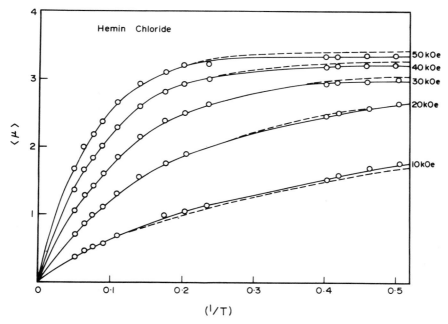

Figure 13 Temperature dependence of the average magnetization of hemin chloride (2–20 K) [44]. The curves are theoretically calculated with $D = 8.0$ cm^{-1}, $J = 0$ (full lines), and $D = 6.9$ cm^{-1}, $ZJ = -0.08$ cm^{-1} (broken lines) [44].

data are well fitted by Equation (1) [or rather to a modification of Eq. (1) including deviation from the axial symmetry] with the D (and E) values as shown in Figure 14. It is interesting that the zero-field splitting in the Fe(PPIXDME) (SC$_6$H$_4$NO$_2$) is much larger than that in the ox—P450$_{cam}$ · S($D = 3.8$ cm^{-1}), suggesting that the difference may be due in part to the protein environment of the prosthetic group.

It may be worth comparing at this stage the magnetic properties of the high-spin ferric porphyrins with those of the corresponding heme proteins. The magnetic properties are in general in close agreement, especially with myoglobin and hemoglobin complexes (Table 3). The temperature dependence of the average magnetic moment and magnetic anisotropy of ferric myoglobin fluoride, for example, is very similar to that of Fe(TTP)Cl. It should however be borne in mind that the magnetic-susceptibility measurements on heme protein are generally not very accurate because of the large and uncertain diamagnetism of the protein, which almost exceeds the paramagnetism of the iron [45]. Methods like Mössbauer and ESR spectroscopy are therefore more useful for the heme proteins. Magnetic studies on metalloporphyrins evidently have an advantage in this respect. Analysis of the ESR data on ferric heme proteins

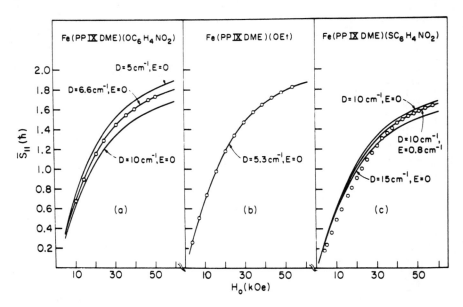

Figure 14 Temperature dependence of the magnetization data for several model iron(III) cytochromes [44a].

shows an ordering of energy levels as in the Fe(TPP)X series [41], with $^4A_2$ being much closer to the ground state than ^4E is. In a tetragonal geometry, as in these heme systems, a point-charge crystal-field model predicts that the elongation will stabilize the d_{z^2} orbital relative to $d_{x^2-y^2}$, and the d_{xz}, d_{yz} pair relative to d_{xy}. The spin quartet state corresponding to the first excited state should then correlate with $(d_{xz}, d_{yz})^3(d_{xy})^1(d_{z^2})^1$, leading to a ^4E term. To explain the observed reversal in the ordering between $^4A_2$ and ^4E, Griffith [41] suggested π-bonding with the heme group, involving electron donation from the filled ligand π-orbital, as a contributing factor.

d. Magnetic Exchange Interaction in High-Spin Iron(III) Porphyrins. Several iron(III) porphyrins show evidence of magnetic exchange interaction in low-temperature magnetic-susceptibility studies [44, 46–52]. This interaction is generally of antiferromagnetic type and can be classified into two groups depending on the structural features of the porphyrins.

To the first group belong a large number of oxobridged binuclear ferric porphyrins in which the interaction is very strong and dominates the magnetic properties at even high temperatures. The most celebrated example of this group is the [FeTPP]$_2$O, with a room-temperature magnetic moment being less than 2 BM. The Mössbauer studies support a high spin state for the ferric ion,

Table 6. Magnetic Exchange Energies in some High-Spin Iron(III) Porphyirins

Compound[a]	D (cm^{-1})	J (cm^{-1})	Model	Ref.
Fe(TPP)$_2$O	[b]	-155	Dimer	[49]
Fe(OEP)$_2$O	[b]	-133	Dimer	[50]
Fe(TPP)$_2$(DuQ)THF	[b]	-15.5	Dimer	[51]
Fe(TPP)$_2 \cdot$ Q $\cdot$ 2THF	[b]	-7.5	Dimer	[51]
Fe(TPP)$_2 \cdot$ (FQ) $\cdot$ THF	[b]	-3.8	Dimer	[51]
Fe(TPP)Cl	[b]	-0.07	Dimer	[21]
Fe(DPDME)Cl	11.0	$ZJ = -0.22$	Mol. field	[44]
Fe(OEP)Cl	8.0	$J = -0.01$	Linear chain	[44]
Fe(PP)Cl, hemin chloride	6.9	$\begin{cases} ZJ = -0.08 \\ J = -0.04 \end{cases}$	Mol. field / Linear chain ($Z = 2$)	[44]

[a] DuQ, Duroquinone; Q, Quinone; FQ, p-Fluoranil.
[b] ZFS not considered in the calculations.

as in the salen compounds. There is some uncertainty as to its room-temperature magnetic moment and magnitude of exchange interaction [47, 52]. A value of $J = 155$ cm^{-1} has been deduced from the NMR studies in solution [49] (see Table 6). A similar value of $J = -133$ cm^{-1} has been deduced for the analogous [Fe(OEP)]$_2$O from a temperature-dependent magnetic-susceptibility study between 1.2 and 293 K [50]. There have been attempts to replace the bridging oxygen by other groups to ascertain its effect on the exchange interaction. In one such attempt a number of dimeric high-spin ferric porphyrins bridged by dianions of hydroquinones has been studied in solid state over a wide range of temperature [51]. These dimers show moderate to weak antiferromagnetic interaction depending on the nature of the dianion (Table 6). A point which deserves mention here is that, in the analysis of the above susceptibility data, a simple $S = \frac{5}{2}$ interacting dimeric model was assumed and the effect of the single-ion zero-field splitting due to spin-orbit coupling of the $^6A_1$ state was not considered. This may be justified in the oxobridged dimers, where the exchange energy is an order of magnitude larger than the zero-field splitting, but it would lead to large errors in those systems where the exchange energy and the zero-field splitting are comparable, as in the hydroquinone bridged ones. A complete theoretical calculation should therefore be used in which both these factors are included.

In the second group are several monomeric iron(III) porphyrins which show small but significant exchange interaction which becomes important at low temperatures. The examples Fe(TPP)Cl, Fe(OEP)Cl, and hemin chloride have been discussed in this context [19, 21, 44, 46]. A striking example is however provided by the magnetization study on chlorodeuteroporphyninato

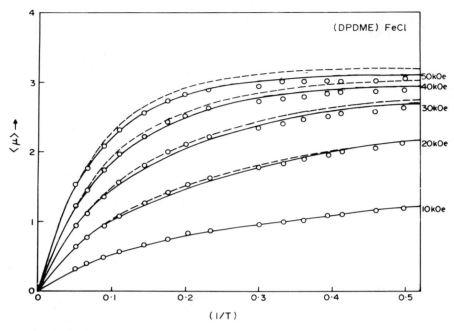

Figure 15 Temperature dependence of the magnetization of ferric deuteroporphyrin dimethylester chloride (2–20 K). The curves are theoretically calculated with $D = 11.0$, $ZJ = -0.22$ cm^{-1} (full lines), and $D = 9.0$, $ZJ = -0.20$ cm^{-1} (broken lines) [44].

iron(III), Fe(DPDME)Cl (Figure 15) [44]. Its magnetization data over the entire magnetic field range could not be fitted with Equation (1) for any single D-value. While the lower-field data (H ≤ 20 kOe) are well fitted by an unreasonably high value of D ($\simeq 30$ cm^{-1}), the high-field data conform well to $D \simeq 11$ cm^{-1}, a value much closer to the far-infrared value, which is 9 cm^{-1} [38]. The average susceptibility data between 4 and 100 K can also be explained only if $D \simeq 30$ cm^{-1}. It has been suggested [44] that the above observations, especially those on the magnetization, indicate a significant antiferromagnetic coupling between the ferric-ion spins in this case. At high magnetic fields this coupling is broken and hence the system appears to behave as a near-normal paramagnet. This can explain why the high-field data are well fitted by a reasonable value of D, while the low-field data are not.

Magnetic exchange interaction in these large bulky molecules must be through the superexchange mechanism. A scrutiny of the packing of these molecules in the lattice reveals the possible pathways for such interactions.

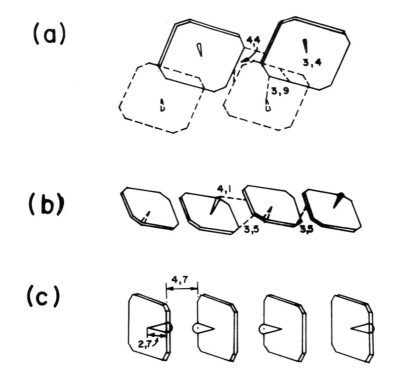

Figure 16 Packing of the porphyrin molecules in (a) hemin chloride, (b) Fe(OEP)Cl, and (c) Fe(TPP)Cl.

This is shown in Figure 16 for Fe(TPP)Cl, hemin chloride, and Fe(OEP)Cl; no structural data are available for Fe(DPDME)Cl. Fe(TPP)Cl may thus be considered as a dimer with only half the molecule involved in the dimer formation, and

$$\text{Fe—Cl} \cdots \text{Cl—Fe}$$

as being the most probable pathway for the magnetic exchange interaction. For hemin chloride and Fe(OEP)Cl the obvious pathway for the superexchange interaction is the "sideways" interaction, the closest intermolecular contacts being 3.39 and 3.54 Å respectively. These contacts however provide a longer pathway through the delocalized macrocyclic porphyrin ring and hence ensure that the interaction energy is very small. It is obvious that the molecular stacking in Fe(DPDME)Cl must be such as to provide a pathway for a much stronger interaction and must be expected to be different from that of the hemin chloride or Fe(OEP)Cl.

The exchange-interaction energy has been calculated accurately in Fe(TPP)Cl by fitting the single-crystal susceptibilities below 4 K with the "dimer" model [21], and a value of $J = -0.07$ cm^{-1} has been deduced. It has been suggested that hemin chloride and Fe(OEP)Cl form a Heisenberg linear chain [46]. The exchange interaction may then be calculated using Fisher's Heisenberg model deduced in the classical-spin limit. A more general way to consider the exchange interaction in these systems, especially in a case like Fe(DPDME)Cl with no structural data, is a molecular-field approximation. The appropriate Hamiltonian, which includes the crystal field and exchange interactions, may be written as [44]

$$\mathcal{H}_s = DS_z^2 + 2\beta \vec{H} \cdot \vec{S} - ZJ\langle \vec{S}\rangle \cdot \vec{S}, \tag{5}$$

where Z is the number of nearest equivalent neighbors interacting with an exchange interaction J, and $\langle \vec{S}\rangle$ is the expectation value of the spin operator $\vec{S}$. Since the value of $\langle \vec{S}\rangle$ depends on $\mathcal{H}_s$, an iterative procedure is used to calculate $\langle \vec{S}\rangle$ self-consistently. The exchange energy deduced by this and linear-chain methods is included in Table 6. For Fe(DPDME)Cl an excellent fit to the entire set of data is now possible with reasonable values of D and ZJ. A very small value of ZJ is required to fit the hemin chloride data to the far-infrared D-value. The sign of ZJ shows that the interaction is antiferromagnetic. The temperature-dependent magnetization data appear to be affected in general by the exchange interaction, and this must be considered in any quantitative analysis of the data. By this criterion the single-crystal magnetic anisotropy has an added advantage in that measurements at even higher temperatures are sensitive to the crystal-field parameters.

From the above discussion it would appear that the magnetic exchange interaction in high-spin iron(III) porphyrins is of rather common occurrence, and is significant in several cases even at high temperatures. This observation need not be true only for the high-spin porphyrins, and indeed should be much more widespread than discussed above. The experimental results on iron porphyrins with other spin states are not still sufficient to display this aspect of their magnetic behavior, and hence will not be discussed further.

iii. Low-Spin Iron(III) Porphyrins

Magnetic studies on low-spin iron(III) porphyrins are extremely limited. Often only room-temperature magnetic moments have been reported for diagnostic purposes; only in a few cases have the measurements even been extended down to 77 K. The magnetic moment at room temperature lies in the expected range of 2.0–2.4 BM and shows only slight temperature dependence. For example, the magnetic moments of Fe(TPP)N$_3$(py) at 297 and 77 K are 2.33 and 2.09 BM respectively [53]. Similar results are obtained for Fe(TPP)Im$_2$(Cl). No single-crystal susceptibility has been reported for any low-spin iron(III) porphyrin, though such data would be very useful. A detailed understanding and analysis of their magnetic properties must await such measurements.

iv. Spin-Mixed Iron(III) Porphyrins

We have examined in Figure 2 the conditions and regions for substantial spin mixing. This region is very narrow in the sextet-doublet crossover range, but is rather significant in the sextet-quartet parameter space, i.e. $36{,}000 \geqslant \Delta_{Oh} \geqslant 31{,}000$ and $C' \geqslant 4{,}000$ cm^{-1}. For $\Delta_{Oh} > 36{,}000$ cm^{-1} a pure spin quartet may exist. Experimental studies on iron(III) porphyrins have so far not been able to establish any example of a *pure* spin-quartet ground state, though several examples are now known where the ground state is supposed to be spin-mixed with substantial spin-quartet or sextet contributions.* Chemical criteria for stabilizing spin quartet ground state have been discussed in Volume I Chapter 2, and it has been argued that an increased equatorial bonding with a weak axial interaction (ClO_4^- etc.) may stabilize at least a predominant spin-quartet ground state [54]. Recently a few iron(III) porphyrins which show anomalous magnetic properties as in Figures 3–5 have been synthesised with ClO_4^- and ethanol as coordinating ligands. They resemble closely the magnetic behavior of some bacterial ferricytochromes.

a. Fe(OEP)(ClO₄) and Fe(TPP)(ClO₄).

Fe(OEP)(ClO$_4$) was originally prepared [55] by reaction of Fe(OEP)Cl with AgClO$_4$ in benzene under reflux, followed by crystallization after filtration of the hot reaction mixture, which gave dark brown needles of Fe(OEP)(ClO$_4$). This procedure is also applicable to preparing Fe(TPP)(ClO$_4$). There is an alternative synthetic route through the cleavage of the μ-oxo derivative with perchloric acid, but the isolation of the desired compound is often complicated by the aqueous medium [56]. For this reason the first method is generally preferred. Both these perchlorates have been isolated and structurally characterized by single-crystal x-ray technique [54, 57]. The iron atom is five-coordinated with the oxygen of the perchlorate occupying the fifth apical position, similarly to the Fe(TPP)X series. The Fe—N$_{pyr}$ and Fe—ct distances (Table 7) in both these perchlorates are much shorter than those in the high-spin [e.g. Fe(TPP)X] series (cf. Table 1). Also the Fe—O bond length is much longer than (e.g.) in the high-spin exchange-coupled (FeTPP)$_2$O, where it is 1.763 Å [58]. There are thus both an increased equatorial interaction and a weaker axial one, with the result that the Δ_{Oh} in these ferric porphyrin perchlorates becomes larger than in the high-spin complexes, pushing the ferric ion towards the spin-mixed region (cf. Figure 2).

---

* Very recently an iron(III) porphyrin with pure $S = \frac{3}{2}$ ground state has been characterised [*Inorg. Chem.* **21** (1982) 1427].

Table 7. X-Ray Structural Data for Perchlorato Iron(III) Porphyrins

Compound	Fe—N (Å)	Fe—O (Å)	Fe—ct (Å)
Fe(OEP)(OClO$_3$)	1.994	2.067	0.26
Fe(TPP)(OClO$_3$)	2.001	2.029	0.28

The magnetic moment of Fe(OEP)(ClO$_4$) was first reported several years ago and found to have a value of 4.8 BM at 288 K [55]. On the basis of this data it was suggested that this compound was in thermal equilibrium between high- and low-spin states. A subsequent measurement [56] of its magnetic susceptibility between 295 and 84 K showed that the magnetic moment decreased from 4.8 BM to 4.1 BM at 84 K. It was contended that this result, together with a temperature-dependent Mössbauer study, ruled out the possibility of a thermal spin equilibrium; instead it was suggested that the results indicate a quantum-mechanical admixed ground state between spin quartet and sextet states, with the spin quartet lying lower. Subsequently an x-ray structural and a magnetic susceptibility study (275–77 K) have also supported this view [57]. The magnetic behavior of Fe(TPP)(ClO$_4$) is very similar, and hence a similar suggestion has been made [54].

Let us now look more closely into the magnetic behavior of these two interesting compounds and examine if the result could be quantitatively explained on the basis of spin-mixed ground states. Very detailed magnetic susceptibility (295–2 K) and magnetisation (2–20 K, 10–50 kOe) measurements have recently been completed [59] on both these perchlorates, and the data tested against the existing theoretical models. The experimental results are summarized in Figures 17 and 18. The close similarity in the detailed magnetic behavior of the two compounds is quite noticeable; the temperature dependences of their magnetic moment and their saturation moments are very similar in the two cases. The rather sharp decrease in the $\bar{\mu}$ at lower temperatures is reminiscent of the sizeable zero-field splitting of the ground state.

It has been suggested [54, 56] that the magnetic behavior of these compounds could be rationalized on the basis of a model due to Maltempo [59a], which is essentially a simplification of the Harris model (Figure 1). Maltempo's model is based on the general observation that for iron(III) porphyrins the $^6A_1$ and $^4A_2$ states come very close together (see Table 5), with all other levels generally lying very high up in energy. Since $^6A_1$ and $^4A_2$ can mix together by spin-orbit coupling, they would substantially modify the magnetism. Maltempo's model has two adjustable parameters, the separation between the $^6A_1$ and $^4A_2$ states (Δ) and the spin-orbit coupling parameter (ζ). In the region of substantial spin mixing these two parameters should be of comparable magnitude. This model gives a good (least-squares) fit to the data, and the parameters derived are included in Table 8. The ground state in both the cases is predominantly of $^4A_2$ character but with substantial mixing of $^6A_1$.

Iron(III) Porphyrins

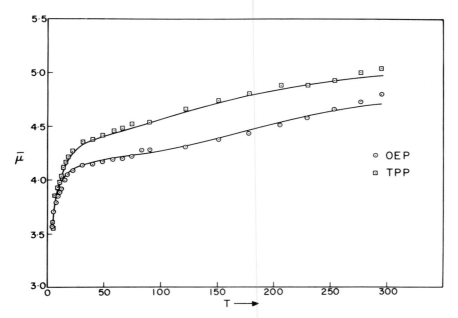

Figure 17 Temperature dependence of the average magnetic susceptibility of the two iron(III) perchlorates [59]. Full curves are theoretically calculated using Maltempo's model (see text).

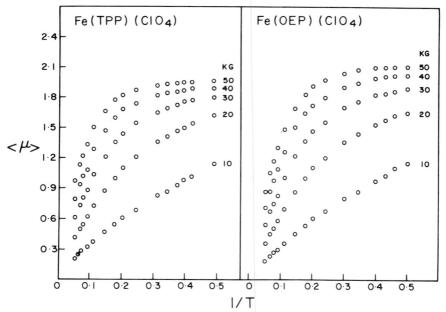

Figure 18 Temperature dependence of magnetization of Fe(TPP)ClO$_4$ and Fe(OEP)ClO$_4$ [59].

Table 8. CF Parameters and ground-state Composition of Iron(III) Perchlorates [59]

Compound	ζ (cm^{-1})	Δ (cm^{-1})	$g_\parallel$	$g_\perp$	Ground state	
Fe(TPP)(ClO$_4$)	188	108	2.00	4.74	$^6A_1$ $^4A_2$	35% 65%
Fe(OEP)(ClO$_4$)	118	204	2.00	4.37	$^6A_1$ $^4A_2$	18% 82%

The derived g-values are also similar to the reported values. The $^4A_2$ state shows large zero-field splitting of about 17 cm^{-1}. The excited $^6A_1$ state appears to be very close, which may have interesting implications.

Even though the fit to Maltempo's model is very good, the spin-orbit coupling parameter derived is rather too low and does not appear to be realistic. This throws some doubt on the validity of this model, and calls for a more detailed calculation. A minor extension of this model would be to include the effect of a ^4E term into Maltempo's model, but this does not appear to have much effect on the magnetic moment [59]. A complete calculation [59] on Harris's model (Figure 1) shows that the mixing of ^4E and the spin doublets into the ground $^4A_2$ is minimal, and it is the mixing with the low-lying $^6A_1$ which accounts for almost the entire magnetism, a situation similar to Maltempo's model.

A further point which is relevant at this stage is to notice that in any of the above variations of the theoretical approaches, the $^4A_2$ and $^6A_1$ come very close together and they would be appreciably populated at room temperature, where the thermal energy is comparable to the separation between the two states. These compounds may then be deemed as showing thermal spin equilibrium together with extensive spin mixing. If the wave functions of the two states contain comparable contributions from these two effects, then the two states may become equivalent as regards the effective spin state, and Mössbauer spectroscopy may fail to show characteristics normally associated with the observation of spin-equilibrium phenomena. A fuller analysis of the magnetic-susceptibility data along these lines may be revealing.

b. Fe(TPP)·2(C$_2$H$_5$OH)(ClO$_4$)·$\frac{1}{2}$CH$_2$Cl$_2$. Recently magnetic and Mössbauer studies on the above compound have been reported, and they offer a good comparison with the compounds discussed earlier [60]. This compound was prepared by treating [Fe(TPP)]$_2$O in dichloromethane with aqueous perchloric acid, and gradually replacing dichloromethane with ethanol. The detailed structure is not known, but is believed that the two ethanol groups may be coordinated as in Fe(TPP)BF$_4$ · 2EtOH. The temperature dependence of the

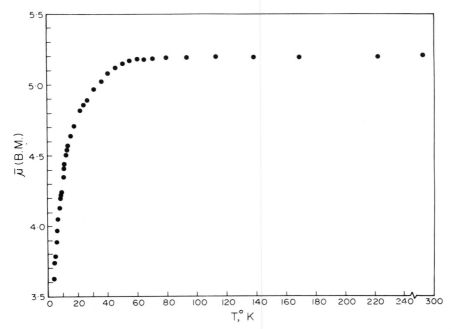

Figure 19 Temperature dependence of the average magnetic moment of Fe(TPP)(ClO$_4$) · 2(C$_2$H$_5$OH) · $\frac{1}{2}$CH$_2$Cl$_2$ [60].

magnetic moment (Figure 19) shows two characteristic features. The magnetic moment at room temperature is about 5.2 BM and remains nearly constant down to $\simeq$ 70 K. The magnetic moment then decreases slowly but steadily, reaching a value of 3.6 BM at 3.5 K. This behavior, despite the similarities at lower temperatures, differs significantly from that of Fe(TPP, OEP)(ClO$_4$) in the variation of the magnetic moment above 70 K. Still more significant is the difference in the quadrupole splitting and its temperature dependence as listed in Table 9. These results have been interpreted to indicate that the ground state of this compound is predominantly of high spin with $S = \frac{3}{2}$ character admixed into it. In this respect it resembles closely the bacterial protein rubrum ferricytochrome c. A detailed analysis of its crystal structure would serve as a useful structural comparison with the predominant $S = \frac{3}{2}$ iron(III) perchlorates.

v. Spin Equilibrium

Spin equilibrium (or spin crossover) in transition-metal complexes is now a well-known (if not equally well-understood) phenomenon. Generally the equilibrium is between two equienergetic spin states (or species), and is very

Table 9. Some Magnetic Data on Spin-Mixed Heme Systems

Compound	Magn. Susc.		Mössbauer			Ref.
	T (K)	$\bar{\mu}$	T (K)	ΔE (mm/sec)	I.S. (mm/sec)	
Fe(TPP)(ClO$_4$)	290	5.05	295	2.79	0.30 ⎱	[54, 59] [a]
	80	4.5	77	3.48	0.38 ⎰	
Fe(OEP)(ClO$_4$)	290	4.78	295	3.16	0.37 ⎱	[56, 57, 59]
	80	4.25	115	3.52	0.37 ⎰	
Fe(TPP)(ClO$_4$)	293	5.2	300	2.04	0.28 ⎱	[60]
2(EtOH)	80	5.2	77	1.98	0.35 ⎰	
Rubrum	290	5.2	4.2	1.35	0.37	[59a]
ferricyto-chrome c	150	5.2				

[a] I.S. = Isomer Shift, ΔE = Quadrupole splitting.

critically dependent on temperature, pressure, and other physical parameters. Factors governing this equilibrium process as well as the magnetic properties expected of such a system near the crossover region have already been discussed in detail [61, 62]. There are several iron heme proteins which exhibit changes characteristic of the spin equilibrium, and show some quite anomalous magnetic behavior [62]. Surprisingly, this phenomenon is not so common in iron(III) porphyrins and has been identified only in a few cases. The best-documented example is a series of derivatives of iron(III) octaethylporphyrin [63], discussed below in some detail.

Magnetic measurements [63] over a range of temperature both in solid and in solution states have been reported on [Fe(OEP)L$_2$]$^+$, where the axial ligand L = Py, 3-ClPy, 4-NH$_2$Py, or 1-MeIm and the counter anion is either ClO$_4^-$ or PF$_6^-$. The results of magnetic measurements on solids are summarized in Figure 20, the solution data being rather similar with minor differences. As expected the 1-MeIm derivative is nearly low-spin and does not show any evidence of spin equilibrium. All other compounds show large and complicated variations of the magnetic moment with temperature and indicate a (high spin) ⇌ (low spin) equilibrium. Interesting differences are observed in the magnetic behavior between the perchlorate and hexafluorophosphate of the same cation, and between complexes with various axial ligands. These variations are however not unexpected for a system in the "crossover" region.

A detailed and in-depth theoretical explanation of these results has recently been given [64] on the basis of the model discussed in Section B (cf. Figure 1). A noninteracting quartet ⇌ doublet equilibrium was found to be inadequate to account for even the room-temperature magnetic moment, especially of the 3-ClPy derivative. A complete calculation based on Figure 1, including the spin-state mixing through spin-orbit coupling, shows that the magnetic properties can be best explained in the region of the parameter space (Figure 2) where

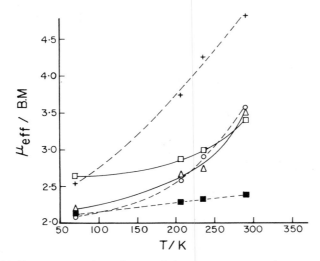

Figure 20 Temperature dependence of the average magnetic moment of several "spin-equilibrium" iron(III) porphyrins [63]: +, [Fe(OEP)(3-ClPy$_2$)]$^+$ PF$_6^-$; □, [Fe(OEP)(Py)$_2$]$^+$ (ClO$_4$); △, [Fe(OEP)(Py)$_2$]$^+$ PF$_6^-$; ○, [Fe(OEP)(4-NH$_2$Py)$_2$]$^+$ PF$_6^-$; ■, [Fe(OEP)(1-MeIm)$_2$]$^+$ PF$_6^-$.

$^6A_1$, $^4A_2$, and ^2E come very close together (i.e., $\Delta_{O_h} \simeq 32{,}000$ and $C' \simeq 4{,}400$ cm^{-1}). However, a unique fit to data particularly those of the 3-ClPy derivative, was found difficult to obtain unless the crystal-field parameters were varied with temperature. The temperature dependence of the parameters was found to be less than $\simeq 3\%$ and is not unlikely in the present situation.

C. IRON(II) PORPHYRINS

Iron(II) porphyrins usually exist in high ($S = 2$) or low ($S = 0$) spin states, but there is at least one well-documented example of intermediate ($S = 1$) spin state. Crystallographic and chemical criteria for stabilizing these spin states have been discussed in Volume I Chapter 2 and are supported by existing experimental evidence. As a sharp contrast to the iron(III) porphyrins, magnetic-susceptibility studies on iron(II) porphyrins are sparse and limited. Measurements on the high-spin iron(II) porphyrins have generally been reported at room temperature, presumably for diagnostic purposes. A detailed magnetic-susceptibility measurement [65] has however been recently reported for the rare $S = 1$ planar FeTPP [66]. This spin state has not as yet been established for heme proteins, though there is some speculation on its existence in certain proteins [67].

We discuss below a ligand-field model [65] for the d^6 (Fe^{2+}) electron configuration describing the conditions for various spin situations, and then illustrate the expected magnetic behavior of the $S = 1$ ground-state complexes with special reference to FeTPP.

i. Theoretical Foundations: The Ligand-Field Model

The electronic structure of iron(II) complexes in lower than octahedral symmetry has been investigated by several workers using ligand-field theory [68, 69]. These calculations refer to limited areas of ligand-field space for potential ground states. We describe below a detailed ligand-field calculation in D_4 symmetry including spin-orbit coupling and apply it to iron(II) porphyrins [65]. The inclusion of spin-orbit coupling in this model causes extensive mixing of states, especially in the crossover regions where there may be several close-lying states.

The ligand-field calculations were performed using ligand-field parameters as effective operators which act upon the d-electron wave functions. The parametrization used was in terms of one-electron energies of the real d-orbitals. Thus for a system in D_4 symmetry there are three parameters representing the relative energies of the a_1 (z^2), b_2 (xy), and b_1 ($x^2 - y^2$) orbitals to the $e(xz, yz)$ set. A complete d^6 basis-set calculation requires the diagonalization of a 210 × 210 matrix and is too large for routine computations. Exclusion of the singlet states reduces the basis set to 160. The computation can be simplified further by using a symmetry-adapted basis set which converts the matrix to be diagonalized into independent blocks corresponding to the irreducible representations of the D_4 group [70, 71].

Figure 21 shows plots of the ground-state energy boundaries as a function of the three energy separations. In this case the parameters for each diagram are the energies of the a_1- and b_2-orbitals with respect to the e-orbitals for different energies of the b_1-orbital. For low values of $E(x^2 - y^2)$ three spin quintets are possible as ground states over a large range of parameter space, ^5E and $^5B_2$ occurring as ground states only when b_2 lies above or below the e-orbital, respectively. As the b_1-orbital moves higher in energy, the probability of $S = 2$ ground states becomes more favored. For $E(x^2 - y^2) = 35,000$ cm^{-1} no $S = 2$ ground state is stabilized, but several possible $S = 1$ ground states now occur in the parameter-space diagram. The regions of parameter space over which the low-spin ($S = 0$) ground state is possible increase with the energy of the b_1 orbital. Thus as the $x^2 - y^2$ orbital moves to much higher energies, either $S = 1$ or 0 becomes the preferred ground state.

We now investigate the magnetic properties of $S = 1$ ground-state systems on the basis of the above model. A general observation with such complexes is that their room-temperature magnetic moment is usually much higher (3.8–4.7 BM) than the spin-only value of 2.89 BM. This high value of the magnetic moment has often misled researchers [69, 72], and a satisfactory explanation of its origin long remained obscure [73]. Calculation of the magnetic susceptibility in the various regions of the parameter space shows that triplet states well

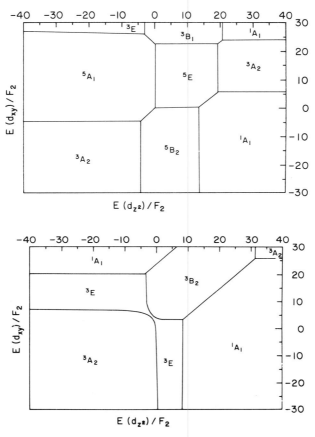

Figure 21 Regions of parameter space defining the different ground states for a d^6 electron configuration in D_4 symmetry: (a) $E(x^2 - y^2) = 15,000$ cm^{-1}, $E(xy, yz) = 0$; (b) $E(x^2 - y^2) = 35,000$ cm^{-1}, $E(xz, yz) = 0$; $F_2 = 1,000$ and $F_4 = 100$ cm^{-1}. Taken from Reference [65].

isolated from each other do not give a sufficiently large magnetic moment. It is only in the crossover region, where the three states $^3A_2$, $^3B_2$, and ^3E (cf. Figure 21) come very close together, that increased magnetic moments as high as 4.8 BM could be found. Figures 22 and 23 show the variation in the average and principal magnetic moments at 300 K in this crossover region. It is immediately apparent that magnetic moments as high as $\simeq 4.9$ BM can be achieved for an $S = 1$ ground-state system in the region where the $[Z^2]$ orbital lies lowest. This observation is quite significant, since complexes with similar high magnetic moments have previously been ascribed to an $S = 2$ ground state [69]. Further, such complexes would show very large paramagnetic anisotropy with $\mu_\perp > \mu_\parallel$ (Figure 23), which is another important characteristic.

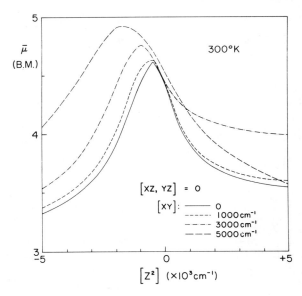

Figure 22 Dependence of the average magnetic moment at 300 K on $E(z^2)$ for various values of $E(xy)$. Here $\zeta = 380$ cm^{-1}; $E(xz, yz) = 0$, $E(x^2 - y^2) = 40,000$ cm^{-1} [65].

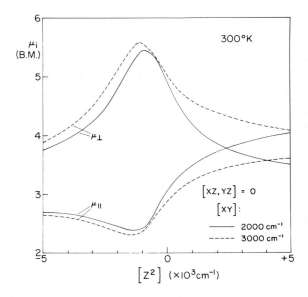

Figure 23 Dependence of principal magnetic moments at 300 K on $E(z^2)$ for different values of $E(xy)$. Here $E(x^2 - y^2) = 40,000$, $\zeta = 380$ cm^{-1}.

Iron(II) Porphyrins

ii. Examples

a. High-Spin Iron(II) Porphyrins. Very few magnetic-susceptibility data exist on the high-spin iron(II) porphyrins. A large number of "picket-fence" iron(II) porphyrins have been shown to have high spin [74], as well as the interesting Fe(TPP)(1-MeIm) derivative [75]. However, no detailed susceptibility study has as yet been reported.

b. $S = 1$ Iron(II) Tetraphenylporphyrin. FeTPP was first prepared [69] by heating Fe(TPP)(Py)$_2$ in vacuum for several hours, which gives a nearly "amorphous"-looking polycrystalline powder (Prep. I). Magnetic measurement on this sample gave a value of 4.75 BM at room temperature. This product is generally very air-sensitive. A subsequent preparation [66] is based on the chromous reduction of Fe(TPP)Cl, which gives well-formed purple tetragonal crystals (Prep. II). The crystals are not as air-sensitive as the previous sample. The magnetic moment of this product is close to 4.2 BM at room temperature [65]. A third method of preparation [76] (Prep. III) is stated to give an air-stable crystalline pure sample with a room-temperature magnetic moment of 4.85 BM. Of all these methods, the sample resulting from Prep. II [i.e. chromous reduction of Fe(TPP)Cl] has been most extensively studied and structurally characterized. The iron is in the plane of the porphyrin macrocycle, with a shorter Fe–N bond [66]. The temperature dependence of its average magnetic moment and magnetic anisotropy is shown in Figures 24 and 25 respectively.

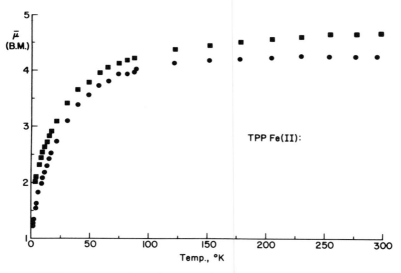

Figure 24 Temperature dependence of the average magnetic moment of FeTPP prepared by two different methods: ■, Prep. I; ●, Prep. II (see text). Data taken from Reference [65].

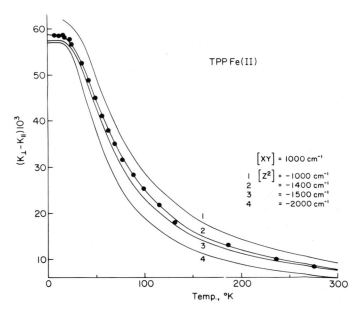

Figure 25 Temperature dependence of the molecular anisotropy of FeTPP. The solid curve is the theoretical fit for $E(x^2 - y^2) = 40{,}000$ cm^{-1}, $E(xy) = 1{,}000$ cm^{-1}, $E(xz, yz) = 0$, $E(z^2) = -1{,}400$ cm^{-1}, $\zeta = 380$. The effect of the change in $E(z^2)$ on the fit is shown [65].

The temperature dependence of $\bar{\mu}$ in Figure 24 bears a close resemblance to that of another $S = 1$ system, iron(II) phthalocyanine (FePc) [77]. The magnetic moment is nearly independent of the temperature above 100 K, below which it decreases with decreasing temperature—first rather slowly, then rapidly. This indicates large zero-field splitting of the ground state. Figure 25 shows that $K_\perp > K_\parallel$ as in FePc, and that the molecular anisotropy below 10 K is nearly independent of temperature. This latter observation is true for the average susceptibility of FeTPP as well [65]. At this stage it is useful to compare the magnetic behavior of FeTPP prepared by the different methods. This is illustrated in Figure 24 for $\bar{\mu}$ for the Prep. I and II. The resemblance is clear—with, however, a subtle difference at higher temperatures (above 100 K), where the Prep. I sample shows an increase in $\bar{\mu}$ with temperature and a significantly higher room-temperature moment. All these results are compatible with the $S = 1$ ground state for FeTPP, irrespective of the preparative method. No detailed magnetic data are however available for Prep. III.

Returning to the results of Figures 24 and 25 (for Prep. II), an attempt was made [65] to fit the data on the basis of a spin-Hamiltonian formalism for

$S = 1$, as has been done for FePc and other compounds [77-78]. The expressions for the principal susceptibilities then become

$$K_{\parallel} = \frac{2N\beta^2}{kT} g_{\parallel}^2 (e^d + 2)^{-1}, \tag{6}$$

$$K_{\perp} = \frac{2N\beta^2}{D} g_{\perp}^2 \frac{e^d - 1}{e^d + 2}, \tag{7}$$

where $d = D/kT$, D being the ZFS parameter.

It has been very convincingly shown that while Equations (6) and (7) reproduce very well the temperature dependence of $\bar{\mu}$ (cf. Figure 24) with $g_{\parallel} = g_{\perp} = 2.94$ and $D = 70$ cm^{-1}, they fail to account for the temperature variation of the molecular anisotropy, suggesting the inapplicability of the spin-Hamiltonian formalism to such systems. This is not surprising in view of the previous observation in Figure 21 that several low-lying levels come close together in such systems. It is however interesting that the average magnetic moment is again quite insensitive to theoretical models. A further point of interest is that Equations (6) and (7) are incompatible with the $\bar{\mu}$-versus-T curve for Prep. I, as they cannot predict a rise in $\bar{\mu}$ with temperature above 100 K.

The ligand-field model gives excellent agreement with both the average susceptibility and molecular anisotropy using the one-electron energies shown in Figure 26. The spin states deduced are also included in the Figure. As discussed earlier, the $[Z^2]$ orbital lies lowest, with $[XZ, YZ]$ and $[XY]$ close to it. Further, Figure 26 shows that there are several low-lying spin states within 1000 cm^{-1}, which explains the reason for the failure of the spin-Hamilton (SH) formalism especially at higher temperatures. The ground state ($^3A_2$) is split by about 90 cm^{-1} (cf. the SH model). The $^3A_2$ state arises from the electron configuration $b_2^2 a_1^2 e^2$ and is the configuration favoured by Mössbauer [66] and NMR [79] studies. Alternative configurations to be considered include $e^4 b_2^1 a_1^1$ and $b_2^2 e^3 a_1^1$, giving $^3B_1$ and ^3E states respectively. It has been shown [65] that these two configurations are close in energy and that terms arising from each of them are extensively mixed by spin-orbit coupling. Hence it is too simplistic to analyze the data in terms of electron configurations; an accurate description would be in terms of spin-mixed states.

It is interesting to compare the ground states of FeTPP and FePc. Mössbauer measurement on crystalline FePc suggests a configuration $b_2^2 e^3 a_1^1$, as against the magnetic-circular-dichroism [80] studies in dichlorobenzene, indicating $b_2^2 e^2 a_1^2$, the same as in FeTPP. This comparison is interesting in view of the difference in the molecular packing in the solid state between FePc and FeTPP. In the FePc crystals, the planar molecules are stacked in the lattice in such a way that the iron may be considered to have virtually a (4 + 2) coordination, with iron atom lying 3.2 Å above or below the nitrogen atoms of the adjacent phthalocyanine molecules. In FeTPP the adjacent molecules are

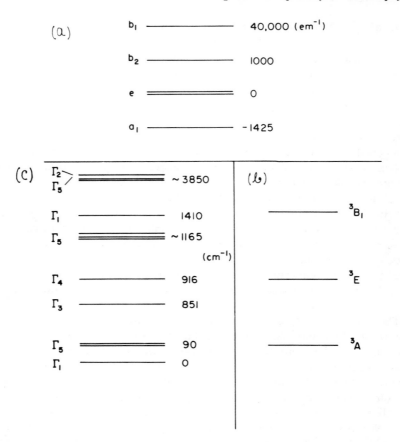

Figure 26 One-electron energies deduced from the anisotropy data (a), terms (b), and spin state (c) in FeTPP [65].

however much too far away to have any effect, and hence the iron remains four-coordinated. In dichloromethane solution the lattice structure is broken and FePc will have a four-coordinated structure, as in FeTPP; hence the similarity in the ground-state configurations.

The magnetic-susceptibility study on FeTPP is thus able to give a wealth of information, particularly since the measurements were carried out with single crystals. This is important for iron(II) porphyrins, since ESR cannot be used for such systems. NMR study [79] has at best been able to make some tentative suggestions regarding the ground-state configuration; no detailed information regarding electronic structure is available. Mössbauer studies, especially at high magnetic field, are useful for the iron(II) porphyrins and can give quite reliable and accurate information if carried out in conjunction with magnetic-susceptibility study.

D. CONCLUDING REMARKS

In the discussions above, we hope we have been able to show the usefulness of the magnetic susceptibility study in understanding the electronic structure of the iron porphyrins. The magnetic studies on these metalloporphyrins have also illustrated similarity with the heme protein in many cases. This observation is significant and makes measurements on metalloporphyrins very important. It is of course recognized that the magnetic-susceptibility measurements must be done over a wide range of temperature, and preferably on single crystals for definitive and detailed informations. We have also illustrated the advantages of low-temperature high-field magnetization studies. With the availability of commercial equipment, measurements over a wide range of temperature and magnetic field is now within easy reach. Single-crystal measurements, though most informative, will remain a specialty and will call for extra effort.

It will also be clear from our discussion in the foregoing sections that magnetic-susceptibility studies on iron porphyrins are still very severely limited. While considerable experimental data exist now on the high-spin iron(III) porphyrins, very little work has been done on the low-spin ones. Only a beginning has been made on the studies on spin-mixed and spin-equilibrium iron(III) porphyrins. Single-crystal susceptibility studies on these systems are very much required. There are virtually no detailed magnetic-susceptibility studies on high-spin iron(II) porphyrins, though similar studies on the corresponding deoxyhemoglobin are available. The success of the studies on FeTPP calls for detailed study of other iron(II) porphyrins.

It is hoped that the present article will be able to convince researchers of the potential of detailed magnetic susceptibility studies on metalloporphyrins and induce them to become involved in these studies.

Acknowledgments

The author is very grateful to Drs. D. V. Behere, R. Birdy, and V. R. Marathe for allowing him to use several unpublished results. The efficient help of the drafting section is also thankfully acknowledged.

References

1. Falk, J. E. *Porphyrins and Metalloporphyrins.* Elsevier, New York, 1964.
2. Smith, K. M. (ed.). *Porphyrins and Metalloporphyrins.* Elsevier, Amsterdam, 1975.
3. Adler, A. D. (ed.). "The Chemical and Physical Behaviour of Porphyrin Compounds and Related Structures" *Ann. N.Y. Acad. Sci.* **206** (1973).
4. Dolphin, D. (ed.). *The Porphyrins*, Academic, New York, 1978, Vols. IIIA, IVB.
5. Collman, J. P. *Accounts Chem. Res.* **10** (1977) 265.
6. Reed, C. A. In *Metal Ions in Biological Systems*, H. Sigel (ed.). Marcel Dekker, New York, 1977, Vol. 7.
7. Scheidt, W. R. *Accounts Chem. Res.* **10** (1977) 339.

8. Mitra, S. *Progress in Inorg. Chem.* **22** (1977) 309.
9. Casey, A. T., and Mitra, S. in *Theory and Applications of Molecular Paramagnetism*, E. A. Boudreaux and L. N. Mulay (eds.). Wiley, New York, 1976.
10. Mitra, S. In *Transition Metal Chemistry*, R. L. Carlin (ed.) Marcel Dekker, New York, 1972, Vol. 7.
11. Lonsdale, K., and Krishnan, K. S. *Proc. Roy. Soc. (London) A* **156** (1936) 597.
12. Smart, J. *Effective Field Theories of Magnetism*. Saunders London, 1966.
13. Harris, G. *Theoret. Chim. Acta*, **10** (1968) 119, 155.
14. Harris, G. *Theoret. Chim. Acta*, **5** (1966) 379.
15. Harris, G. *J. Chem. Phys.* **48** (1968) 2191.
16. Kotani, M. *Progr. Theoret. Phys. (Kyoto) Suppl.* **4**, No. 17, (1961) 1; *Adv. Chem. Phys.* **7** (1964) 159.
17. Marathe, V. R., and Mitra, S. *Indian J. Pure Appl. Phys.* **14** (1976) 893.
18. Maricondi, C., Swift, W., and Straub, D. K. *J. Am. Chem. Soc.* **91** (1969) 5205.
19. Behere, D. V. and Mitra, S. *Indian J. Chem.*, **19A** (1980) 505.
20. Behere, D. V., Date, S. K., and Mitra, S. *Chem. Phys. Letters* **68** (1979) 544.
21. Neiheisel, G. L., Imes, J. L., and Pratt, W. P., Jr. *Phys. Rev. Letters* **35** (1975) 101.
22. Behere, D. V., Marathe, V. R., and Mitra, S. *J. Am. Chem. Soc.* **99** (1977) 4149.
23. Behere, D. V., and Mitra, S. *Inorg. Chem.* **18** (1979) 1723.
24. Behere, D. V., Birdy, R., and Mitra, S. *Inorg. Chem.* **20** (1981) 2786.
25. Maricondi, C., Straub, D. K., and Epstein, L. M. *J. Am. Chem. Soc.* **94** (1972) 4157.
26. Dolphin, D., Sams, J. R., Tsin, T. B., and Wong, K. L. *J. Am. Chem. Soc.* **100** (1979) 1711.
27. Venoyama, H. *Biochim. Biophys. Acta* **230** (1971) 479.
28. Venoyama, H., and Sakai, K. *Spectrochim. Acta* **31A** (1975) 1517.
29. Sato, M., Rispin, A. S., and Kon, H. *Chem. Phys.* **18** (1976) 211.
30. La Mar, G. N., Eaton, G. R., Holm, R. H., and Walker, F. A. *J. Am. Chem. Soc.* **95** (1973) 63.
31. Walker, F. A., and La Mar, G. N. *Ann. N.Y. Acad. Sci.* **206** (1973) 328.
32. Behere, D. V., Birdy, R., and Mitra, S. *Inorg. Chem.* **21** (1982) 386.
33. Loew, G. H., *J. Mag. Reson.* **6** (1972) 408.
34. Marathe, V. R., and Mitra, S. *Chem. Phys. Letters* **19** (1973) 140.
35. Gerloch, M., Lewis, J., and Slade, R. C. *J. Chem. Soc. (A)*, 1969, p. 1442.
36. Marathe, V. R., and Mitra, S. *Chem. Phys. Letters* **27** (1974) 103.
37. Vermaas, H., and Groneveld, W. L. *Chem. Phys. Letters* **27** (1974) 583.
38. Brackett, G. C., Richards, P. L., and Caughey, W. S. *J. Chem. Phys.* **54** (1971) 4383.
39. Venoyama, H., Iizyka, T., Morimoto, H., and Kotani, M. *Biochem. Biophys. Acta*, **160** (1968) 159.
40. Birdy, R., Behere, D. V., and Mitra, S. *J. Chem. Phys.* (Feb. 1983 issue).
41. Griffith, J. S., *Proc. Roy. Soc. (London) A* **235** (1956) 23; *Nature* **180** (1957) 30.
42. Koenig, D. F., *Acta Cryst.* **18** (1965) 663.
43. Sullivan, S., Hambright, P., Evans, B. J., Thorpe, A., and Weaver, J. A. *Arch. Biochim. Biophys.* **137** (1970) 51.

44. Marathe, V. R., and Mitra, S. *J. Chem. Phys.* (Jan. 1983 issue).
44a. Tang, S. C., Koch, S., Papnefihymiou, G. C., Foner, S., Frankel, R. B., Ibers, J., and Holm, R. H. *J. Am. Chem. Soc.* **98** (1976) 2415.
45. Tasaki, A., Otsuka, J., and Kotani, M. *Biochim. Biophys. Acta*, **140** (1967) 284.
46. Ernst, J., Subramanian, J., and Fuhrhop, J. H. *Z. Naturforsch.* **32a** (1977) 1129.
47. Fleisher, E. B., and Srivastava, T. S. *J. Am. Chem. Soc.* **91** (1969) 2403.
48. Cohen, I. A. *J. Am. Chem. Soc.* **91** (1969) 1980.
49. Boyd, P. D. W., and Smith, T. D. *Inorg. Chem.* **10** (1971) 2041.
50. Moss, T. H., Lilienthal, H. R., Moleski, C., Smythe, G. A., McDaniel, M. C., and Caughey, W. S. *J. Chem. Soc. Chem. Comm.* 1972, p. 263.
51. Kessel, S. L., and Hendrickson, D. N. *Inorg. Chem.* **19** (1980) 1883.
52. Cohen, I. A. *Structure and Bonding* **40** (1980) 1.
53. Adams, K. M., Rasmussen, P. G., Scheidt, W. R., and Hatano, K. *Inorg. Chem.* **18** (1979) 1892.
54. Reed, C. A., Mashiko, T., Bentley, S. P., Kastner, M. E., Scheidt, W. R., Spartalian, K., and Lang, G. *J. Am. Chem. Soc.* **101** (1979) 2948.
55. Ogoshi, H., Watanabe, E., and Yoshida, Z. *Chem. Letters* 1973, p. 989.
56. Dolphin, D. H., Sams, J. R., and Tsin, T. B. *Inorg. Chem.* **16** (1977) 711.
57. Masuda, H., Taga, T., Osaki, K., Sugimoto, H., Yoshida, Z., and Ogoshi, H. *Inorg. Chem.*, **19** (1980) 950.
58. Hoffman, A. B., Collins, D. M., Day, V. W., Fleischer, F. B., Srivastva, T. S., and Hoard, J. L. *J. Am. Chem. Soc.* **94** (1972) 3620.
59. Mitra, S., Marathe, V. R., and Birdy, R. *Chem. Phys. Letters*, **90** (1982) to appear.
59a. Maltempo, M. M., *J. Chem. Phys.* **61** (1974) 2540.
60. Mitra, S., Date, S. K., Nipankar, S. V., Birdy, R., and Girerd, J. J. *Proc. Indian Acad. (Chem. Sci.)* **89** (1980) 511.
61. Martin, R. L., and White, A. H. In *Transition Metal Chemistry*, R. L. Carlin, (ed.). Marcel Dekker, New York, 1968, Vol. 4.
62. George, P., Beetlestone, J., and Griffith, J. S. *Rev. Mod. Phys.*, **36** (1964) 441.
63. Hill, H. A. O., Skyte, P. D., Buchler, J. W., Lueken, H., Tonn, M., Gregson, A. K., and Pellizer, G. *J. Chem. Soc. Chem. Comm.*, 1979, p. 151.
64. Gregson, A. K. *Inorg. Chem.* **20** (1981) 81.
65. Boyd, P. D. W., Buckingham, D. A., McMeeking, R. F., and Mitra, S. *Inorg. Chem.* **18** (1979) 3585.
66. Collman, J. P., Hoard, J. L., Kim, N., Lang, G., and Reed, C. A. *J. Am. Chem. Soc.* **97** (1975) 2676.
67. Maxwell, J. C., and Caughey, W. S. *Biochemistry* **15** (1976) 388.
68. Konig, E., and Schnakig, R. *Theor. Chim. Acta* **30** (1973) 205.
69. Kobayashi, H., and Yanagawa, Y. *Bull. Chem. Soc. Japan* **45** (1972) 450.
70. Griffith, J. S. *The Theory of Transition Metal Ions*, Cambridge U. P., New York, 1964.
71. Gerloch, M., and McMeeking, R. F. *J. Chem. Soc. Dalton, Trans.*, 1976, p. 2443.
72. Dezsi, I., Belazs, A., Molnar, B., Gorobchenko, V. D., and Lukashevich, I. I. *J. Inorg. Nucl. Chem.* **31** (1969) 1661.
73. Figgis, B. N., *Introduction to Ligand Fields*, Interscience, New York, 1966, Chapter 12.

74. Collman, J. P., Gagne, R. R., Reed, C. A., Halbert, T. R., Lang, G., and Robinson, W. T. *J. Am. Chem. Soc.* **97** (1975) 1427.
75. Collman, J. P., and Reed, C. A. *J. Am. Chem. Soc.* **95** (1973) 2048.
76. Hussain, S. M., and Jones, J. G., *Inorg. Nucl. Chem. Lett.* **10** (1974) 105.
77. Barraclough, C. G., Martin, R. L., Mitra, S., and Sherwood, R. C. *J. Chem. Phys.* **53** (1970) 1643.
78. Prins, R., and van Voorst, J. D. W. *Chem. Phys. Letters* **1** (1967) 54.
79. Goff, H., La Mar, G. N., and Reed, C. A. *J. Am. Chem. Soc.* **99** (1977) 3641.
80. Stillman, M. J., and Thomson, A. J. *J. Chem. Soc. Faraday Trans. II* **70** (1974) 790.

2
Electron Paramagnetic Resonance of Hemoproteins

GRAHAM PALMER

A. INTRODUCTION

In an earlier chapter in Volume I Scheidt and Gouterman (Chapter 2) have reviewed in detail the variety of oxidation and spin states available to heme iron. They noted that at least four oxidation states ($+1 - +4$) are experimentally accessible, although only the $+2$ and $+3$ states are commonly encountered. Each of these oxidation states can exist in several spin states, encompassing all possibilities between $S = 0$ and $S = \frac{5}{2}$.

Despite this range of oxidation- and spin-state alternatives, only two species are routinely observed by electron paramagnetic resonance spectroscopy (EPR). These are the low-spin ($S = \frac{1}{2}$) and high-spin ($S = \frac{5}{2}$) configurations available to ferric iron, and almost all published data on the EPR of heme iron have been obtained on these two species. Consequently, the bulk of this chapter will be devoted to describing the properties of these two states of Fe^{3+}.

The balance of the chapter will deal with some specialized systems: the EPR of $S = \frac{3}{2}$ heme and of $S = \frac{3}{2} - \frac{5}{2}$ quantum mixed states, the EPR of the $S = \frac{1}{2}$ spin state of the "Fe^{4+}" center of the peroxidase family enzymes, and finally, the EPR properties of heme-nitrosyl compounds. Although this latter topic does not deal literally with the electronic properties of the iron atom, the EPR spectra of heme-NO complexes have been of considerable value in addressing structural and functional questions in several hemoprotein systems, and it is appropriate that a review of these compounds be included in this chapter.

While ligand hyperfine contributions are often observed in the EPR of transition-metal ions, it is a fact that they are rarely to be observed in hemeprotein spectra. The magnitude of such interactions, particularly those from the nitrogens of pyrrole and imidazole, have been investigated using electron-nuclear double resonance (ENDOR) spectroscopy [1], but this technique is not routinely available to the experimentalist. A summary of all but the most recent results has been published [2], and this topic will not be considered in detail in this chapter.

i. Relevant EPR Concepts

Electron-paramagnetic-resonance spectroscopy probes the magnetic moment (μ) associated with the unpaired electron(s) present in a paramagnetic heme:

$$\mu^2 = \bar{g}^2 S(S + 1), \qquad (1)$$

where $2S$ is the number of electrons present in the paramagnet, and $\bar{g}$, the g-factor, is a parameter which specifies the strength of interaction of the spin system with an applied magnetic field H_0. For $S = \frac{1}{2}$ systems g is a direct measure of the size of the magnetic moment:

$$\bar{g}^2 = \tfrac{1}{3}\left(g_x^2 + g_y^2 + g_z^2\right). \tag{2}$$

However, if $S > \tfrac{1}{2}$, then the observed g-value is usually not a direct measure of the total spin S, but of a unique subset of the total spin manifold, a concept which will become clearer after we have considered the EPR properties of the high-spin Fe^{3+} configuration. In real compounds, $\bar{g}$ depends upon the coordination number and geometry of the paramagnetic center, and the chemistry of the primary ligands to the heme.

The g-factor is in fact the most obvious piece of information to be extracted from an EPR spectrum. Experimentally it is defined as

$$g = h\nu/\beta H_0, \tag{3}$$

where h and β are Planck's constant and the Bohr magneton respectively, ν is the operating frequency of the spectrometer used (typically about 9 GHz), and H_0 is the value of the applied magnetic field used to locate a specific feature in the EPR spectrum. However, this familiar relationship obscures the essential fact that the g-factor is a tensor and the value of g obtained using Equation (3) will depend directly on the orientation of the paramagnetic species with respect to the direction of the applied magnetic field.

This concept can be illustrated explicitly for cytochrome c, where, for the purposes of this illustration, we make the simplifying approximation that the principal axes of the g-tensor of the ferric heme coincide precisely with a rectangular coordinate system fixed on the heme structure (Figure 1).

When a crystal of cytochrome c is placed in the applied magnetic field and oriented so that the field is perpendicular to the heme plane (i.e. parallel to the heme normal), then, for a spectrometer operating at 9.2 GHz, the value of H_0 necessary to obtain maximum EPR absorption is 2148 gauss. Rotating the crystal so that $\vec{H}_0$ now lies along the axis connecting pyrroles II and IV (Figure 1) shifts the field necessary for maximum absorption to 2934 gauss. A further rotation which aligns $\vec{H}_0$ with the axis connecting pyrroles I and III moves the resonant field to 5300 gauss. Thus, depending upon the orientation of a molecule of cytochrome c, we would calculate g-values of 1.45, 2.22, or 3.06. These three values are the principal g-values for cytochrome c; they are most commonly referred to as g_x, g_y, and g_z respectively.* The relationship between the directions of the components of the g-tensor and the structure of the heme implied in this example and specified by the subscripts to the components of g

---

* Alternative nomenclatures are available, e.g.: g_1, g_2, and g_3, or g_{min}, g_{mid}, and g_{max}. This last system appears the most attractive because it contains no implicit assumptions, and simply alludes to the relative magnitudes of the three values. However it seems probable that the convention used in the text will continue to dominate the literature, despite the implicit, and frequently unjustified, correlation of structural and magnetic axes.

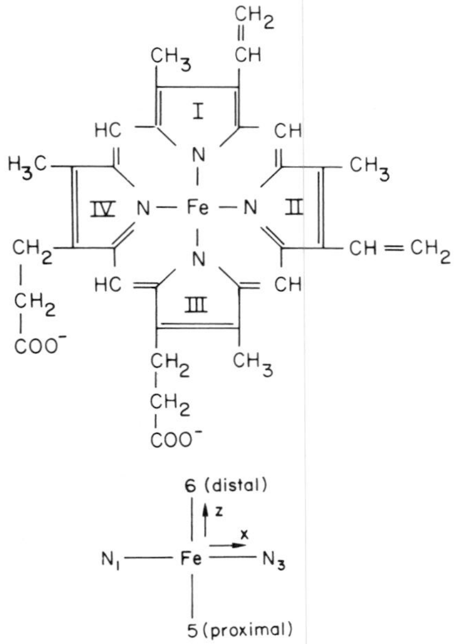

Introduction

Figure 1 Top: The structure of protoheme, identifying the standard nomenclature (I–IV) for the pyrrole rings. Bottom: Edge-on view of the heme, revealing the proximal and distal coordination sites.

will be referred to frequently in this chapter. For simplicity therefore, I will refer to this set of coordinates as "our usual coordinate system".

In practice, few data have been obtained on single crystals, and most experiments are conducted on so-called "powder samples" such as frozen, aqueous solutions of proteins or, for model studies, low-temperature glasses of the model compound in a suitable organic solvent. Thus the sample placed in the EPR spectrometer typically contains about 10^{16} molecules; these are presumed to be randomly oriented. Certainly, experimental spectra can be well simulated with this assumption, implying that the cooling and freezing of the sample do not lead to any obvious orientation of the molecules in the sample.

In these powder samples, a given molecule exhibits a g-value defined by

$$g_{obs}^2 = g_x^2 l_x^2 + g_y^2 l_y^2 + g_z^2 l_z^2, \qquad (4)$$

where l_x^2, l_y^2, and l_z^2 are the direction cosines defining the projection of the applied magnetic field onto the x, y, and z axes of the g-tensor. Thus $g_x^2 l_x^2$, $g_y^2 l_y^2$, and $g_z^2 l_z^2$ are the contributions of the three principal g-values to g_{obs}, the g-factor exhibited by an arbitrarily oriented molecule. As the orientation changes, so do l_x^2, l_y^2, and l_z^2, and thus each orientation is associated with its characteristic g-value.

A small number of molecules will have $\vec{H}_0$ parallel to x; then $l_x^2 = 1$, $l_y^2 = l_z^2 = 0$, and these molecules will exhibit g_x; similarly, molecules which lie with y or z parallel to $\vec{H}_0$ will exhibit g_y or g_z respectively. However, most of the molecules in the sample will lie with their principal axes cocked with respect to $\vec{H}_0$, and these will exhibit a g-value intermediate between the two extremes, g_x and g_z; thus for cytochrome c, EPR absorption sets in at 2148 gauss and extends to 5300 gauss. However, the EPR intensity at intermediate values of H_0 does not have a constant amplitude, for when $l_x^2(g_x^2 - g_y^2) = -l_z^2(g_z^2 - g_y^2)$, $g_{\text{obs}} = g_y$. Thus the middle g-value is exhibited not only by y-oriented molecules, but also by molecules which satisfy the relationship just expressed, with the important consequence that the EPR intensity rises to a maximum at the magnetic field corresponding to g_y.

This is illustrated in Figure 2(c), which shows, by the broken line, a typical plot of the number of paramagnetic centers per unit field (dN/dH) as a function of H_0. The solid line in Figure 2(c) shows the result of adding a real line shape to this distribution. Because of the design of the detection circuits in the EPR spectrometer, the experimental presentation is the first derivative of

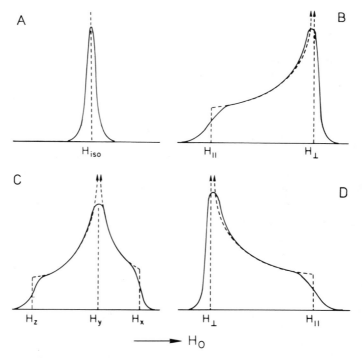

Figure 2 Idealized and real EPR absorption envelopes for the standard anisotropies: (a) isotropic, $g_x = g_y = g_z$, (b) axial, $g_\parallel > g_\perp$, (c) Rhombic, $g_x \neq g_y \neq g_z$, (d) axial, $g_\parallel < g_\perp$.

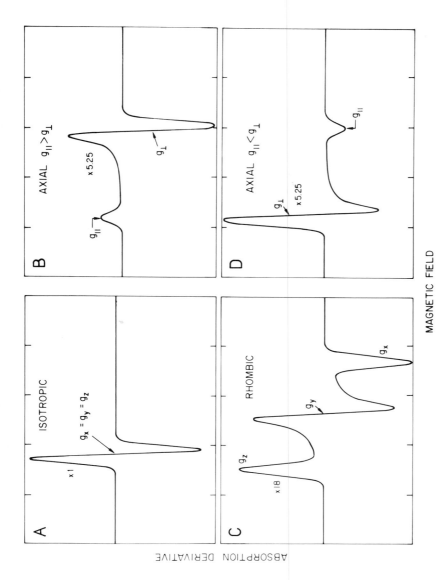

Figure 3 First-derivative lineshapes of the absorption curves shown in Figure 2: (a) isotropic, (b) axial, $g_{\parallel} > g_{\perp}$, (c) rhombic, (d) axial, $g_{\parallel} < g_{\perp}$. After Palmer, G., in *Methods for Determining Metal Ion Environments in Proteins*, D. W. Darnell and R. C. Wilkins (eds.), Elsevier North Holland, New York, 1980, pp. 153–182.

this distribution function; the corresponding lineshape is shown in Figure 3(c). An explicit example is provided by the EPR spectrum of cytochrome c (Figure 4).

The existence of three unequal g-values is referred to as the rhombic case and implies that the microsymmetry about the heme is no higher than rhombic. Symmetries lower than rhombic cannot be established by inspection of EPR spectrum, but can be deduced from a study of the relative orientations of the g- and A-tensors, or from a study of the linear electric-field effect.

When two of the g-values are numerically equal, the resulting EPR spectrum is said to be axially symmetric (implying an axially symmetric microgeometry for the paramagnet). The unique g-value is called $g_\parallel$ (parallel to the symmetry axis, $= g_z$), and the other two g-values are referred to as $g_\perp$. Two cases exist:

(a) $g_\parallel < g_\perp$ [e.g., high-spin ferric heme; see Figures 2(d), 3(d)]. An example is aquo-metmyoglobin (Figure 5).
(b) $g_\parallel > g_\perp$ [e.g., Cu(II); see Figures 2(b), 3(b)].

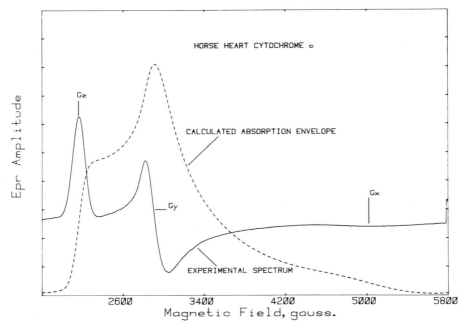

Figure 4 Experimental EPR spectrum of horse-heart cytochrome c recorded at 10 K. The solid line shows the direct spectometer output with the g-values indicated (the frequency was 9.24 GHz). The dashed line shows the absorption envelope obtained by numerical integration. Note the extremely small amplitude with associated large width at g_x. (G. Palmer, unpublished results).

Introduction

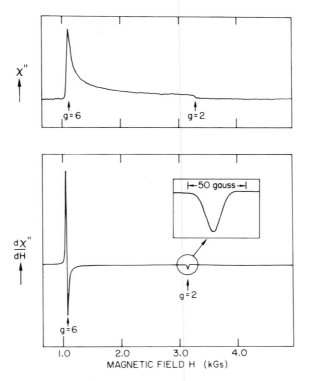

Figure 5 Experimental EPR spectrum of aquo-metmyoglobin. Top: Experimental absorption curve obtained by temperature modulation of the sample. Bottom: Conventional first-derivative spectrum obtained by magnetic-field modulation of the sample [after Feher, G., Isaacson, R. A., Scholes, C. P., and Nagel, R., *Ann. N.Y. Acad. Sci.* **222** (1972) 86–102].

Finally, when $g_x = g_y = g_z$ we have the isotropic case [Figures 2(a), 3(a)]. Equation (2) is normally written in polar coordinates using the identities

$$l_x^2 = \sin^2\theta \cos^2\phi,$$
$$l_y^2 = \sin^2\theta \sin^2\phi, \qquad (5)$$
$$l_z^2 = \cos^2\theta,$$

where θ is the polar angle between z and $\vec{H}_0$, and ϕ measures the amount of rotation between the x-axis and the projection of $\vec{H}_0$ onto the xy plane. Then

$$g_{obs}^2 = \left(g_x^2\cos^2\phi + g_y^2\sin^2\phi\right)\sin^2\theta + g_z^2\cos^2\theta. \qquad (6)$$

This form of the equation is the one typically employed in the simulation of the EPR spectra of powder samples.

B. EPR OF LOW-SPIN FERRI-HEMOPROTEINS

The electronic configuration of a low-spin ferric hemoprotein is usually written as $(t_{2g})^5$ or explicitly as $(d_{xy}^2 d_{xz}^2 d_{yz})$. Possible contributions from $d_{x^2-y^2}$ and d_{z^2} are presumed to be small, and are ignored in most treatments. Were the microsymmetry at the heme perfectly octahedral, then the three t_{2g} orbitals would have equal energy, the unpaired electron would be distributed equally among all three wave functions, and the electron spin would reside in a spatially spherical environment. There would be no unique axis for developing a net orbital angular momentum, and consequently the g-factor would be established solely by the electron spin with $g_x = g_y = g_z = 2.0$. Conversely, if the unpaired spin were confined completely to d_{yz} and the other t_{2g} orbitals were (energetically) so far removed that they could not be mixed into d_{yz} by spin-orbit interaction, then again there could be no net orbital angular momentum, and all g-values would be two.

However, the well-known observation of three unequal g-values in these systems (e.g. Figure 4) demonstrates clearly that the three t_{2g} orbitals reside close together, but are of unequal energy (Figure 6), so that the unpaired electron acquires orbital contributions to the magnetic moment by the mixing together of the three t_{2g} orbitals.* This inequivalence is brought about by an asymmetric ligand field due principally to the character of the two ligand functions present in the fifth and sixth coordination sites on the heme iron; in cytochrome c, these functions are provided by N_3 of histidine 18 and the thioether S of methionine 80.

The nonspherical component of the ligand field can be decomposed into two components. The larger axial component is z-directed and destabilizes d_{xz} and d_{yz} with respect to d_{xy} by an amount Δ (Figure 6). The second, rhombic component produces an inequivalence in the x- and y-directions such that d_{xz} is rendered more stable than d_{yz} by an amount V (Figure 6). The relative stabilizations of d_{yz}, d_{xz}, and d_{xy} are 0, A, and B. In a popular convention V/Δ is never larger than $\frac{2}{3}$, for in the limit of $V/\Delta = \frac{2}{3}$ the three wave functions are separated by equal amounts (see Figure 6). A larger value of V leads to a qualitative change in Figure 6 such that it is possible to relabel the crystal-field axes to restore the requirement $V \leq \frac{2}{3}\Delta$. An axis system which conforms to this convention is called a proper axis system [3]. When $V = \frac{2}{3}\Delta$, the ligand field is said to be maximally rhombic. The molecular basis for this asymmetry in the ligand field will be considered later.

---

* At this point it might be useful to note two conventions prevalent in the description of this system. First, the symbols ξ, η, and ζ are frequently used in place of d_{yz}, d_{xz}, and d_{xy}. Secondly, calculations are frequently implemented using the "hole" formalism in which the order of the t_{2g} functions (Figure 6) is inverted and a "hole" (positive electron) placed in the lowest state, now d_{yz}. The sign of the spin-orbit constant is reversed to accommodate the change in the charge of the spin.

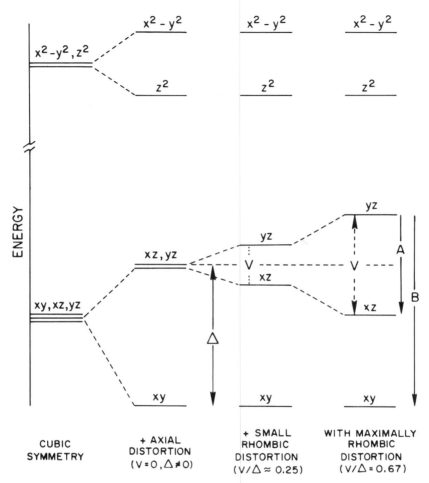

Figure 6 Energy levels of the five $3d$ wave functions of Fe^{3+}, showing the splitting pattern of the t_{2g} orbitals as the symmetry is reduced from cubic. The energy parameters V, Δ, A, and B are indicated.

The analysis of the anisotropic EPR of low-spin t_{2g} systems is due primarily to Griffith [4] and Kotani [5], and has been described in detail by Weissbluth [6] and by Palmer [2]. The approach is fraught with convention [2], but Taylor [7], with his "magic method", has exploited a particular representation of the wave function that leads to a convenient simplification of the analysis. He writes [7]

$$|+\rangle = a|yz^+\rangle - ib|xz^+\rangle - c|xy^-\rangle,$$
$$|-\rangle = -a|yz^-\rangle - ib|xz^-\rangle - c|xy^+\rangle. \quad (7)$$

$|+\rangle$ is the composite wave function containing the spin-up (+) electron, $|yz^+\rangle$ is d_{yz}, $i = \sqrt{-1}$, and a, b, and c are real, signed numbers such that $a^2 + b^2 + c^2 = 1$; b and c are approximately $\lambda/2A$ and $\lambda/2B$ respectively, where λ is the spin-orbit coupling constant. These latter "identities" reveal that it is the combined action of spin-orbit coupling and the asymmetric ligand field that establishes the composition of the ground-state wave function.

Taylor's representation of the paramagnetic wave function leads to the following useful expressions [7, 8]:

$$g_z = 2[(a+b)^2 - c^2], \tag{8a}$$

$$g_y = 2[(a+c)^2 - b^2], \tag{8b}$$

$$g_x = 2[a^2 - (b+c)^2], \tag{8c}$$

$$a = \frac{g_z + g_y}{D}, \tag{8d}$$

$$b = \frac{g_z - g_x}{D}, \tag{8e}$$

$$c = \frac{g_y - g_x}{D}, \tag{8f}$$

$$D = [8(g_z + g_y - g_x)]^{1/2}, \tag{8g}$$

$$A = E(yz) - E(xz) = \frac{a+b}{2b} = \frac{g_x}{g_z + g_y} + \frac{g_y}{g_z - g_x},$$

$$B = E(yz) - E(xy) = \frac{a+b}{2c} = \frac{g_x}{g_z + g_y} + \frac{g_z}{g_z - g_x}, \tag{8h}$$

$$\frac{V}{\lambda} = A, \tag{8i}$$

$$\frac{D}{\lambda} = B - \frac{A}{2} = \frac{g_x}{2(g_z + g_y)} + \frac{g_z}{g_y - g_x} - \frac{g_y}{2(g_z - g_x)},$$

$$g_x^2 + g_y^2 + g_z^2 + g_{yz} - g_{xz} - g_{yx} - 4(g_z + g_y - g_x) = 0, \tag{8j}$$

$$g_z + g_y = g_x = 2(a + b + c)^2. \tag{8k}$$

The relationships (8a)–(8c) substantiate the remarks made at the beginning of this section regarding the g-values to be expected for an isolated orbital ($a = 1$, $b = c = 0$) and for a triply degenerate set ($a = b = c = 0.577$). In both cases we see that the magnitude of all three g-values is 2.0.

Note that the definitions of the two energy splittings A and B are given in units of the spin-orbit coupling constant λ. This constant has the value of about 420 cm^{-1} for Fe (free ion). Thus the solution to the equations (8g)–(8i) yields only the relative separation of the three d-orbitals; absolute values for the splittings can only be obtained if the value of λ appropriate for the specific molecule under study is available. This is rarely the case, because λ is sensitive to the degree of covalency of the paramagnetic center and decreases in magnitude as the covalency increases. For example, the value for ferricyanide is about 280 cm^{-1}. Fortunately it is the relative values of A, B, V, and Δ that are useful for our purposes. Note also that the requirement that $a^2 + b^2 + c^2 = 1$ leads to Equation (8j), which makes it clear that knowledge of any two g-values immediately establishes the missing value. This relationship is quite useful when studying the more anisotropic systems, for when g_x is small the associated EPR feature is usually very broad and consequently of extremely low amplitude (see below). Thus, Equation (8k) can be used to aid in searching for a weak feature at high field or, in the worst case, in providing a numerical value for the missing g-value.

i. The Interpretation of Low-Spin Ferric Heme EPR Spectra

However, the most popular use of the above relationships is in the exploitation of the low-spin heme correlation diagrams (Figure 7) introduced by Blumberg and Peisach [9]. In these diagrams, the energy parameters V and Δ are exploited as indicators of the strength and asymmetry of the ligand field exerted by the two apical ligands at the central metal ion. These two ligands are most commonly provided by side chains of the polypeptide, but it is often possible to append an externally added reagent (e.g. CN, N_3) to one of the apical sites (Figure 1). Blumberg and Peisach measured the g-values of a very large number of naturally occurring hemeproteins, together with compounds derived from them by reaction with exogenous ligands, and appropriate model compounds. They subsequently calculated the values of V/λ and D/λ for each species. These values were then recorded on a plot of V/Δ versus Δ/λ (see Figure 7). In this plot, the dimensionless x-axis provides a measure of the strength of the axial ligand field sensed by the Fe orbitals; thus compounds located along the same vertical line will experience the same relative charge, but will not necessarily be of the same geometry. The ordinate measures the in-plane asymmetry produced by the axial ligands; it varies from pure axial ($V/\Delta = 0$) to purely rhombic ($V/\Delta = \frac{2}{3}$, in a proper coordinate system). Thus hemeproteins located on the same horizontal line experience the same relative asymmetry, but will be subjected to varying amounts of axially directed charge.

It is important to note that in the original usage of these diagrams, Blumberg and Peisach [9] did not use the coordinate system we have adopted, but selected an alternative set of axes which can be derived from our choice by the transformations $x \to -z$, $y \to -x$, $z \to y$; this alternative is perfectly legitimate because it satisfies two conditions believed to be true for the low-spin d^5-systems, viz.: the quantities $g_x g_y g_z$ and $g_x + g_y - g_z$ are both

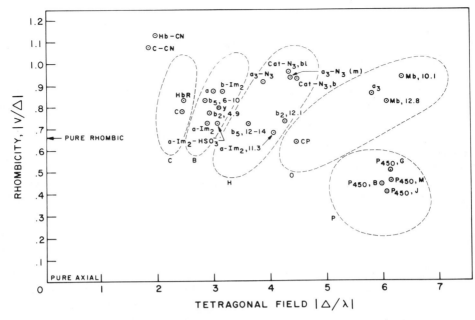

Figure 7 Correlation diagram for low-spin hemoproteins. Modified from Blumberg and Peisach [9] to show horse and yeast cytochrome a (a and y respectively) and bis-imidazole protoheme [23].

positive. (In principle, the g-product can have either sign; the positive sign has been assumed in the past and confirmed recently for several hemoproteins from analysis of Mössbauer spectra [10].) Because this latter axis system does not restrict V/Δ to the range $0-\frac{2}{3}$, it is called an improper axis system, although this should not be taken to imply that it is illegal, or less valid than a proper system.

Blumberg and Peisach discovered that the results from a large body of data fell into five major domains which they labelled C, B, H, O, and P. The first four of these were centered at about the same value of V/Δ, while moving systematically to higher values of Δ/λ. All of these proteins have the imidazole of histidine as the proximal (fifth) heme ligand (Figure 1) and the microenvironment about the heme only obviously differs by virtue of the character of the group present in the distal (sixth) coordination site. It is this latter ligand which is primarily responsible for the observed variations in the g-values, and the existence of these domains suggests that proteins that cluster in the same domain have similar groups present in the distal site. The domains were identified as arising from thio-ether sulfur (C), imidazole N—NH (H), imidazole N—N (B), and OH (O) (the distinction between the H and B

domains depends upon the state of protonation of nonbonding nitrogen of one of the two histidines).

The fifth (P) group appears to be unique, with the smallest value of V/Δ and the largest value for Δ/λ. Blumberg and Peisach [9] believe that this zone is characteristic of the ligand pair imidazole-thiol (i.e., histidine plus cysteine) [11]. They noted that the crystal-field parameters for both microsomal and bacterial P450 all fall within the same range as several model systems with the presumed appropriate ligand pair—for example, hemoglobin Kansas plus mercaptoethanol and a small-molecular-weight species consisting of thiol:heme:imidazole. By contrast, thiol:heme:amine, ethoxide:heme: imidazole, and a presumed thiol:heme:H_2O complex all yielded crystal-field parameters which were clearly different. However, complexes of thiols with hemoglobin and myoglobin fell into a zone somewhat shifted with respect to that assigned to the P450 family, in obvious contradiction to the conclusions most simply drawn from the other data. The suggested resolution of this difficulty proposes a thiolate:heme:ImH ligand system for the first (P450) family and the thiolate:heme:Im$^-$ configuration for the model compounds of hemoglobin and myoglobin.

The presence of sulfur as a heme ligand in the P450 series is supported by a variety of other spectroscopic data. Thus, there is an intense vibrational band at 351 cm^{-1} in the resonance Raman spectrum of P450 [12]. Although originally interpreted as an Fe—N mode, it is now recognized that the iron-imidazole stretch falls within the range 220–290 cm^{-1} [13] while Fe-pyrrole skeletal modes are very weak. The observed line at 351 cm^{-1} is much more plausibly interpreted as an Fe—S mode, by analogy to both the iron-sulfur proteins [14] and to heme halide model compounds [15]. Likewise, analysis of the EXAFS scattering at the Fe of P450 leads to the conclusion that there is a heavy atom, presumably S, within the coordination system of the heme iron [16]. Finally, the detailed interpretation [17] of the unusual optical properties of the heme chromophore in the high-spin P450-camphor complex and the diamagnetic ferrous P450-CO complex requires the presence of a strong ligand akin to sulfur at one apical site.

There are two additional facts that need to be included in this review. First, Griffin and Peterson [18] find that P450$_{cam}$ strongly enhances the relaxation rate of H_2O, a result that was interpreted as indicating that protons can enter the first coordination shell of the iron. This result has been confirmed by ENDOR measurements [19], which reveal the presence of a strongly coupled, exchangeable proton near the heme iron. From the magnitude of the hyperfine coupling constants a structure of the type Fe—X—H was deduced, where X could be either O, N, or S. However, it was concluded that the N_1 proton of a histidine coordinating iron through N_3 would, at 5 Å, be too far away to account for the observed hyperfine interaction. A second observation made in this ENDOR study was the lack of any features attributable to the nitrogen nuclei of histidine in the spectrum of the high-spin P450-camphor complex. As such features are commonly observed in the ENDOR spectra of high-spin hemeproteins, this observation suggests that histidine is not a ligand to iron in this species.

A second important aspect of the coordination properties of P450 is the observation that the EPR spectrum of the reduced P450-NO complex exhibits only a three-line hyperfine-splitting pattern [20]. Thus, in this derivative the ligand trans to the NO does not have a nuclear moment; the suggestion that this trans ligand might still be a histidine but with an abnormally long Fe—N bond is not supported by calculation [21], which shows that the magnitude of the coupling of the unpaired electron to the trans nucleus should be essentially independent of bond length (the use of NO as a heme probe will be described in the last section of this chapter). The possibility that addition of NO ruptures the bond between the iron and the trans ligand is ruled out from the g-anisotropy of the spectrum, which showed the iron had maintained 6-coordination [20]. The nitric oxide data then are quite consistent with the formulation S—Fe—NO for the reduced protein, and the overall picture that emerges is

$$S—Fe(III)—XH \text{ (low-spin)} + NO + e^- \rightarrow S—Fe(II)—NO. \quad (9)$$

It would thus seem prudent to conclude that the ligand pair to be associated with the heme in the P450 family is not defined at this time; sulfur is almost certainly the ligating atom at one site, but the nature of the amino acid occupying the second site has yet to be reliably established.

A straightforward example of the use of the low-spin correlation diagrams can be drawn from recent work on cytochrome oxidase. One of the heme centers (cytochrome a) present in this enxzyme is low-spin, with g-values of 1.24, 2.24, and 3.03 [22]. Direct use of Equations (8f) and (8g) yields values of 4.4 and 0.5 for Δ/λ and V/Δ respectively. These values locate the heme close to the middle of the B-domain (Figure 7), implying that the ligands to the heme in this cytochrome are a pair of histidine residues. This conclusion is reinforced when bona fide bis-imidazole heme a is examined; its crystal-field parameters are 4.8 and 0.52, which also falls within the B-domain [23].

However, in making correlations of this kind, there is an unspoken assumption that all of the species that are being compared have the same absolute orientation for the g-tensor. Regrettably, this information is rarely available, requiring spectroscopic measurements on crystals or some other form of oriented sample. In the case of cytochrome oxidase it is known that partially dehydrated films are highly oriented in two dimensions [24]. A comparison of the magnetic and optical anisotropy of this film established that g_z is perpendicular to the heme plane [25], which suggests that the requirement of the coincidence of the magnetic axes will be met in this case.

It is instructive to apply the same procedure to P450. For example, the g-values for the protein from phenobarbital-induced rat-liver microsomes are

1.922, 2.243, and 2.416 [11]. If we assume our usual orientation for these g-values, this is an improper axis system for P450. However, the choice $g_x = -2.243$, $g_y = 2.416$, and $g_z = -1.932$ leads to values of -6.45 and -0.46 for the axial field and rhombicity.* This latter choice does conform to the requirements of a proper axis system; the associated values of a, b, and c [Equation (7)] are 0.106, 0.069, and 0.996, so that the unpaired electron now resides in d_{xy} (g_x and g_y are the largest g-values). Note however that our choice of axes is specified by the directions of the g-tensor. For the information to be interpreted in terms of the heme coordinate system, we need to know the relationship between the two axis systems, and only if the g-value along the heme normal were found to be 1.932 would the magnetic and structural axes be related. Alternatively put, the orientation of orbital containing unpaired electron deduced from the g-values is the plane in which g-values of 2.243 and 2.416 are measured. Only if this plane coincides with the plane of the porphyrin ring will "d_{yz}" have the orientation that we usually depict. If, in fact, future experiment shows that the g-value along the heme normal is 2.416, then our usual coordinate system, although improper in the formal sense, will be more meaningful for our needs.

ii. Highly Anisotropic Low-Spin Hemeproteins

The g-values for g_z that we have so far encountered have been restricted to the range 2.3–3.1. These species are much the more readily observed, but it should be stressed that much larger values for g_z are in fact quite common. The most extreme value appears to be 3.76; it is observed in cytochrome b_T (b_{568}) of the mitochondrial electron-transfer chain. Some representative data on these more anisotropic species are summarized in Table 1. Clearly a variety of interesting and important proteins are to be found in this category.

The most striking aspect of this table is the relatively few entries for g_x and g_y. There are several reasons for this circumstance. The spectra of these species are extremely temperature-dependent. The low-field peak, being the sharpest feature in the spectrum, is the most readily visualized, whereas the derivative-like middle feature, and especially the high-field trough, are typically extremely broad and often can only be observed with confidence when the operating temperature is extremely low, e.g. 1.4–4.2 K. Indeed, it has been noted that these g-values are better determined from EPR absorption spectra obtained under rapid-passage conditions [26]. The difficulty in observing the middle g-value is compounded in systems which contain more than one paramagnetic species, for the overlap of the several slightly separated derivative line shapes leads to an overall envelope that lacks well-defined extrema.

---

*As already noted, this choice of axes meets the requirements that $g_x g_y g_z$ and $g_z + g_y - g_x$ be positive.

Table 1. EPR Parameters for Selected Highly Anisotropic Low-Spin Hemes

Compound (source)	g_x	g_y	g_z [a]	Ref.	Chirality [b]	Ref.
c-type cytochromes						
c (heart)	1.24	2.24	3.06	[29]	R	[30]
c_{552} (Euglena)	?	?	?		R	[31]
c_{558} (Euglena)	1.39	2.05	3.20	[26]	n.a.	
c_{551} (Ps. aeruginosa)	(1.25)	2.05	3.2	[32]	S	[31]
c_{551} (Ps. stutzeri)	?	?	?		S	[33]
c_{557} (Crithidia oncopeltii)	?	?	?		R	[34]
c_2 (Rhodospirillum rubrum)	1.23	2.11	3.13	[26]	n.a.	
c_{550} (Paracoccus denitrificans)	?	2.06	3.27	[26]	n.a.	
c_1 (heart)	?	?	3.44	[35]	n.a.	
c_1 (yeast)	?	?	3.50	[36]	n.a.	
Other cytochromes						
Liver microsomal b_5	1.41	2.22	3.05	[37]		
Cytochrome f (spinach)	(1.6)	2.07	3.48	[38]		
Cytochrome b_{562} (heart)	?	?	3.44	[35]		
Cytochrome b_{562} (yeast)	?	?	3.60	[36]		
Cytochrome b_{568} (heart)	?	3.713 3.715		[35]		
Cytochrome b_{568} (yeast)	?	?	3.76	[36]		
Model compounds						
Cytochrome c · cyanide	0.93	1.89	3.45	[26]		
Alkaline cytochrome c (his—Fe—NH2)	1.50	2.06	3.35	[26]		
Metmyoglobin · pyridine	1.17	2.19	3.14	[39]		
PPIX · bis(pyridine) [c]	< 0.8	1.35	3.53	[39]		
PP · bis(pyrrole)	< 0.8	0.9	3.7	[39]		
PPIX · bis(butylamine)	?	?	3.6	[26]		

[a] ? = values not reported.
[b] See text for definition.
[c] PPIX = protoheme IX.

The ligand fields that lead to these highly anisotropic g-tensors are somewhat speculative, there being very few relevant model systems available. However the observations on bis-butylamine heme [26] and the presumed cytochrome c–amine system [26, 27] led to the proposal that lysine is a coordinating ligand in these proteins [25]. Thus, it has been suggested that g_z-values of about 3.4 arise from the NH_2:Fe:Im configuration, while the largest values for g_z are due to the NH_2:Fe:NH_2 ligand system (Table 1). Further discrimination of the observed g-values obtained by variations in the state of protonation of the histidine is quite speculative, and much more model work is needed on species which exhibit these patterns of g-anisotropy.

For example, we have recently found that the low-field g-value produced in bis-imidazole heme complexes is very sensitive to the steric strain imposed by the ligand. Thus, the complex of protoheme with 1-methylimidazole has a low-field g-value of 3.00 and exhibits a nicely resolved rhombic EPR spectrum leading to crystal-field parameters typical of the B-domain. By contrast, complexes of protoheme with 2-methylimidazole and 1,2-dimethylimidazole exhibit low-field g-values close to 3.5, while g_y and g_x are either poorly defined or not observable [28]. It thus becomes apparent that the crystal-field parameters V and Δ can be significantly perturbed by steric effects, which presumably lead to a lengthening of the Fe—imidazole bond. Most of the data upon which the Blumberg-Peisach diagrams are founded were obtained with complexes in which the ligands were at their equilibrium distance. This additional aspect of these interactions partially undermines the usefulness of these correlations and suggests for example, that the mitochondrial cytochrome(s) b, may still be bis-histidine derivatives.

iii. Which Way Is "y"?

By and large, we now recognize that the paramagnetic electron in these systems is located in an orbital which, to a first approximation, can be identified as d_{yz}. The orientation of the z-axis has been known for a long time from EPR studies on single crystals; it lies essentially along the heme normal (deviations of up to 20° are found in practice).

But where in the porphyrin plane do the x- and y-axes lie? In principle, one also obtains this information from the single-crystal measurements, but the interpretation of the data usually suffers from a twofold uncertainty that precludes the unambiguous assignment of the x- and y-axes [2]. Fortunately, NMR provides a means of establishing this orientation [40]. Using the crystallographically determined spatial locations of the local polypeptide protons, it is possible to calculate the dipolar contribution to the NMR paramagnetic shift for each proton for any orientation of d_{yz}. These calculated values can then be compared with the experimental data and the orientation of d_{yz} adjusted to provide the closest agreement with observation. For two proteins, cytochromes b_5 [40] and c [41], it has been reported that good agreement is only obtained if d_{yz} is aligned close to the bisector of pyrroles II and IV (Figure 1), as I had implied at the beginning of this chapter.

We may then ask what structural feature dictates this orientation. The most obvious explanation invokes an anisotropic interaction between the metal d-orbitals and the π-orbitals on one or both of the axial ligands, which can destabilize d_{yz} with respect to d_{xz}. The filled $p\pi$ orbital of the imidazole would be a suitable vehicle for this purpose, as it lies parallel to the heme plane. However, in cytochrome c, the plane of the proximal histidine coincides with the axis connecting the α and γ methine bridges (Figure 1), leaving this ligand symmetrically located with respect to the two $d\pi$ orbitals. In this geometry, the ligand $p\pi$ orbital interacts equally with the two metal orbitals, and thus cannot be the source of the interaction, which renders them energetically inequivalent.

Wuthrich and his colleagues [42] have proposed that it is the orientation of the $\varepsilon\text{-}CH_3$ of the ligating methionine which provides the necessary factor (Figure 8). They base their proposal on two pieces of NMR data. First, the orientation of g_y in cytochrome c_{551} from *Pseudomonas aeruginosa* falls along the bisector of pyrroles I and III [43], i.e., in this protein, the x and y coordinates are rotated 90° with respect to the heart protein. Secondly, the paramagnetic shifts observed for the $\varepsilon\text{-}CH_3$ resonance of the methionine require that in the heart protein, the methyl group be located roughly over the nitrogen of pyrrole I (the R-conformation; see Figure 8), whereas in the bacterial protein the methyl group is located close to the nitrogen atom of pyrrole IV (the S-conformation). Thus the π-bonding lone pair on the sulfur atom points along the II-IV axis in the heart protein and along the I-III axis in the bacterial cytochrome. The difference in electronic interaction between this lone pair and the two $d\pi$ orbitals raises the energy of the metal orbital aligned with the sulfur lone pair; this destabilized orbital is the recipient of the paramagnetic electron, and is consequently identified as d_{yz}.

With only two exceptions, the chirality of the methionine conformation observed in the heart protein has been found in all c-type cytochromes examined (Table 1); the only other example of the R-conformation is the protein from *P. stutzeri*. Although there are only a few data available (Table 1), it does appear that there is no correlation between these two conformational variants of cytochrome c and the two sets of EPR g-values reported by Brautigan et al. [26], who noted that the g-tensors of c-type cytochromes clustered into two groups, typified by the values exhibited by the horse and Euglena proteins (Table 1).

In cytochrome b_5, the situation is a little more confusing. The plane of histidine 39 falls along a line connecting methine β with methine δ, and thus

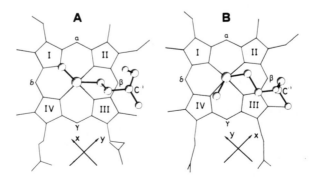

Figure 8 The two conformations of methionine in c-type cytochromes. Parts (a) and (b) show the R and S conformations respectively. Reproduced from [42].

the lone pair of this imidazole interacts equally with the two $d\pi$ orbitals. Histidine 63, however, lies approximately along the II-IV axis, previously reported as the orientation of g_y [40]. In this case, the filled $p\pi$ orbital falls along the I-II axis, and should raise the energy of the d-orbital oriented in the same direction. Thus, the obvious prediction has g_y lying along the I-III axis, in contradiction to its reported orientation along the II-IV direction.

A resolution of this inconsistency can be found in the recent report [44] that the published heme orientation in cytochrome b_5 may be in error. From high-resolution NMR measurements, it was deduced that the correct orientation of the heme differs from that originally established by a 180° rotation about the axis connecting the a and γ methine bridges. Because of the nature of the analysis used in establishing the direction of g_y, this proposed revision in the heme orientation should lead to an interchange in the directions originally assigned to x and y, while the locations of the histidines would be unaffected. The role of histidine 63 in establishing the orientation of g_y would then be quite plausible.

iv. Consequences of Field-Swept Spectra

The measurement of EPR intensity is an important aspect of experimental EPR spectroscopy, for the absolute quantity of an observed paramagnetic species present in a sample colors one's opinions about the significance of that species in the system being studied. In performing such intensity measurements, it is necessary to obtain the area under the EPR absorption curve. In practice, this is done by measuring the amplitude of the spectrum as a function of magnetic field at fixed increments, dH, of the independent variable H_0, but, as Aasa and Vanngard recently pointed out [45], the theoretical expression for the intensity [46] is derived as a function of energy ($=$ frequency $\equiv g$) and thus the integration is more properly performed using equal increments of energy as the independent variable, for example by dividing the spectrum into equal increments of dg rather than dH. The integrals obtained by the two methods are simply related:

$$f(g) = f'(H) \cdot g', \tag{10}$$

where g' is a simple function of the individual g-values [45]. It was this revision in accounting procedures that allowed Aasa et al. [22] to establish that the low-spin component of cytochrome oxidase, cytochrome a, comprised 1 (mole heme)/(mole enzyme). Previous estimates of this species had been consistently low by about 30%, a result that had complicated the understanding of the nature of the heme centers in this enzyme [47].

A second consequence of field-swept spectra is to be found in the overall appearance of the EPR spectra, in particular the striking difference in line-width which is observed at the extreme ends of the spectra (Figure 4).

It is a common observation in the EPR spectroscopy of metalloproteins that the widths of the individual features vary with the operating frequency of the spectrometer. This behavior is most strikingly seen in the iron-sulfur proteins. With these proteins, the spectral linewidths are almost linearly dependent on the spectrometer frequency [48]. The conventional explanation for this behavior invokes a phenomenon called "g-strain". This phenomenon asserts that the three principal g-values do not have unique values, but must be represented by a distribution. This distribution arises from microheterogeneity in the structural parameters of the paramagnetic center, which leads to variations in the energy terms V and Δ (which should depend on bond lengths, etc.). As the g-values depend explicitly on these energy terms, they will reflect this distribution. Experimentally, this distribution in g is observed as a contribution to the linewidth that is field-dependent (in contrast, for example, to the unresolved hyperfine interactions, which lead to a field-independent contribution to the linewidth). However, the g-values are reciprocally related to the magnetic field and thus this distribution in g will contribute unequally to the width at the two ends of the spectrum, with the low-field region being much less affected than the high-field region. For example, the measured width of the low-field peak of cytochrome c is about 140 gauss at X-band (Figure 4); this translates into a δg of about 0.2. Assuming the same g-distribution in all regions of the spectrum, one predicts a width of about 900 gauss at the high-field trough; the experimental width is a little tricky to establish (Figure 4), but is totally consistent with this calculated value, lying between 900 and 1000 gauss. Furthermore, the spectral envelope at g_x will be quite asymmetric, with the high-field side of the trough being some 25% broader than the low-field side. It is clear that in cytochrome c, and presumably in most other systems, the striking variations in linewidth and lineshape reflect simply the g-strain present at the paramagnetic center, and contain no obvious geometric information. Hagen has recently examined these points in detail [49].

The existence of the complications just described raises the issue whether or not the EPR of highly anisotropic $S = \frac{1}{2}$ spectra should be presented on a linear g-value scale. The increasing integration of EPR spectrometers with computers make this a straightforward operation, and such presentation would unquestionably facilitate the evaluation of spectral intensities and the appreciation of intrinsic linewidth anisotropies present in the data.

C. EPR OF HIGH-SPIN HEMEPROTEINS

Just as the EPR of the low-spin hemoproteins is dictated by the symmetry of the ligand field at the iron atom, we also find that the EPR characteristics of the high-spin (^6A) species is a direct consequence of the less than cubic symmetry that exists at the iron center when coordinated by the porphyrin plus axial ligand(s). The electronic configuration of the iron is $t_{2g}^3 e_g^2$ with one electron in each of the five $3d$ wave functions. The total spin is $\frac{5}{2}$, but this can be attained in more than thirty alternative ways by permuting the spin-up ($+$) and spin-down ($-$) electrons singly amongst the five orbitals [e.g., $(+++++)$, $\mathscr{A}(-++++)$, $(--+++)$]. In strict octahedral symmetry, these alternative configurations are equi-energetic in the absence of a magnetic field. When the field is increased, these states diverge in energy with a dependence on H_0 that is linearly proportional to the net number of unpaired electrons (M_s), i.e., proportional to 5, 3, and 1 in the explicit examples just given. Because these levels are split equally, one would observe a single, isotropic resonance at $g = 2.0$.

As we have already seen, in real compounds the symmetry is lowered, with the possibility of both axial and rhombic contributions to the ligand fields. Axial distortions, now represented by the symbol D ($=\Delta$), lead to a separation of states proportional to the magnitude of M_s (Figure 9) so that the $\pm\frac{1}{2}$, $\pm\frac{3}{2}$, and $\pm\frac{5}{2}$ configurations are separated in energy from each other even in the absence of H_0. The rhombic contribution to the ligand field, now identified by the symbol E ($=V/2$), produces a splitting of states with the same absolute value for M_s (Figure 9). These splittings are produced once again by spin-orbit coupling, which in this instance mixes various excited states (predominantly $^4T_2$) into the ^6A ground state.

In cubic symmetry, such mixing alters the center of gravity of all six M_s-states without destroying their equivalence. However, in an axial ligand field ^4T is split into ^4A and ^4E (Figure 9). As spin-orbit coupling between states of different total spin will only mix together substates of the same M_s-values, the effect of the spin-orbit interaction is to stabilize the $M_s = \pm\frac{1}{2}$ and $\pm\frac{3}{2}$ states with respect to the $\pm\frac{5}{2}$ states, with the $\pm\frac{1}{2}$ states being stabilized by the greatest amount. These relative changes are usually written parametrically in the form:

$$E = D[M_s^2 - \tfrac{1}{3}S(S+1)], \tag{11}$$

where the term in S defines the new center of gravity produced by the interaction and D, the so-called zero-field splitting parameter, is given by

$$D = \frac{\lambda}{5}\left(\frac{1}{\Delta_1} - \frac{1}{\Delta_2}\right). \tag{12}$$

Δ_1 and Δ_2 are the excitation energies to the ^4A and ^4E states respectively (Figure 9). In heme compounds Δ_1 is smaller than Δ_2, D is positive, and the relative energies of the three pairs of M_s-states are $-6D$, $-4D$, and 0 for the $\frac{1}{2}$, $\frac{3}{2}$, and $\frac{5}{2}$ doublets, respectively.

Electron spin resonance is only observed from the lowest of the three sets of M_s-doublets. The observed g-values are

$$\begin{aligned} g_\perp &= 3g_e, \\ g_\parallel &= g_e. \end{aligned} \tag{13}$$

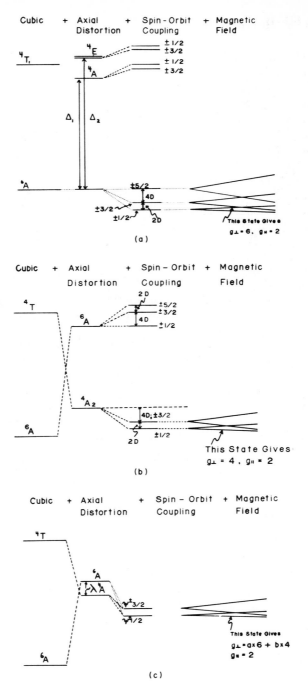

Figure 9 The crystal-field splitting patterns for intermediate- and high-spin Fe^{+3}. (a) Δ/λ large and negative, (b) Δ/λ large and positive, (c) $\Delta/\lambda \simeq 1$. (After Maltempo [58].)

Thus, with $\vec{H}_0$ lying along the heme axis, a g-value of 2.0 is obtained. However, with $\vec{H}_0$ lying in the heme plane, the measured g-value is 6.0. This unusual result can be understood as follows [50]. Normally, the application of a magnetic field to an ensemble of electrons aligns the unpaired electrons along the field direction regardless of the orientation of the molecule. However, in the presence of the strong electron correlation implied by the zero-field splitting, the spins align along the symmetry axis of the asymmetric ligand field responsible for enabling this interaction. This symmetry axis is established by the geometry of the molecule and lies along the heme normal in this example. When this electron-electron interaction is sufficiently large, application of the magnetic field no longer aligns the electrons along the field direction; rather the electrons remain parallel to the heme normal at all times. Thus, when the magnetic field is aligned along the heme normal, (i.e. parallel to the zero-field splitting axis), the magnetic field sees the normal projection of the spin magnetic moment and $g = 2.0$. However, when the magnetic field lies in the porphyrin plane, it is perpendicular to the total spin magnetic moment, and the projection of the magnetic moment onto the field is abnormally large. The magnetic field thus sees an unusually large magnetic moment, which in turn leads to a g-value much larger than normal; specifically, it is three times larger.

The magnitude of D can be obtained in a number of ways: (1) from the temperature dependence of the EPR intensity [51], (2) from the dependence of the magnitude of $g_\perp$ upon the intensity of the magnetic field under conditions where D and the Zeeman interaction are comparable in magnitude [52]; (3) from variable-temperature magnetic-susceptibility measurements [53]; and (4) from far-infrared spectroscopy [54].

Using one or more of these methods, the value of D has been measured for a number of hemoproteins. Typical values fall in the range 4–15 cm^{-1}, and thus the zero-field splitting is very much larger than the Zeeman interaction, which is about 0.1–1 cm^{-1} under experimental conditions normally accessible in the laboratory. Furthermore, all high-spin hemes that experience only axial ligand fields exhibit essentially identical EPR spectra (Figure 5), a result that is somewhat distressing to the experimentalist.

The situation improves somewhat when there are rhombic contributions to the ligand field. As we have seen, these low-symmetry fields modify the energies of d_{xy} and d_{yz} (and also of d_{z^2} and $d_{x^2-y^2}$), and this leads to a splitting of the ^4E excited state into E_x and E_y (Figure 10). The consequences of the resultant spin-orbit interaction are then more complicated, and is usually written

$$D[S_z^2 - \tfrac{1}{3}S(S + 1)] + E(S_x^2 - S_y^2), \tag{14}$$

revealing that there are a further corrections to the energy when the field is in the plane of the heme. For small values of E/D (< 0.1) the expressions for the

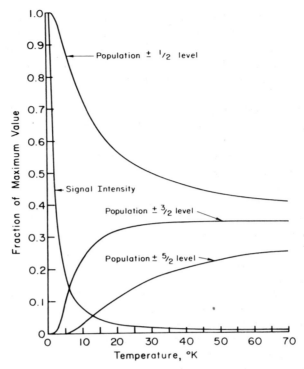

Figure 10 Temperature dependence of the population of the three Kramers doublets ($\pm \frac{1}{2}, \pm \frac{3}{2}, \pm \frac{5}{2}$) and of the signal intensity associated with the $\pm \frac{1}{2}$ component. The figure represents the behavior of the high-spin form of P450$_{cam}$, which has g-values at $\simeq 8$, 4, and 1.8. The energy separations are: $\pm \frac{1}{2}$ to $\pm \frac{3}{2}$, 11.7°; $\pm \frac{1}{2}$ to $\pm \frac{5}{2}$, 33.3°. These differ from the ratio 2 : 6 because the rhombic zero-field parameter $\neq 0$. Data of J. Peisach and W. Blumberg, published in Orme-Johnson and Orme-Johnson, *Methods in Enzymology* **52** (1978) 252–275.

g-values are (55)

$$g_{x,y} = 3g_e \pm 24\frac{E}{D} - \frac{168}{9}\frac{E^2}{D^2},$$

$$g_z = g_e - \frac{304}{9}\frac{E^2}{D^2},$$

(15)

with D defined as before (Δ_2 is now the mean of Δ_x and Δ_y, the excitation energies to E_x and E_y) and

$$E = \frac{\lambda^2}{10}\left(\frac{1}{\Delta_x} - \frac{1}{\Delta_y}\right).$$

(16)

Typically E is small; for example, one of the largest cases of a distorted high-spin heme is provided by P450, for which E is about $0.1D$.

The geometry of these low-symmetry centers can be defined in terms of their rhombicity. From the equations for the g-values just given, we find that $g_y - g_x = 48E/D$, so that, in a proper coordinate system—with $E/D < \frac{1}{3}$ ($\equiv V/D < \frac{2}{3}$)—we have $g_y - g_x < 16$. The rhombicity expressed as a percentage is then $6.25(g_y - g_x)$. For P450 $g_x = 3.7$, $g_y = 8.0$, so that $\Delta_g = 4.3$ and the rhombicity is 27%. This is an extremely large value, most proteins exhibiting relatively little rhombicity, though it should be appreciated that small values are rather difficult to obtain reliably because of the problems associated with extracting g-values from spectra that are only marginally distorted away from axial symmetry.

There are two important practical aspects to the EPR of high-spin heme. The first is illustrated in Figure 10, which shows how the populations of the three Kramers doublets vary with increasing temperature. Of relevance is the decrease in the population of the EPR-detectable $\pm \frac{1}{2}$ state, which drops off exponentially as the temperature approaches and surpasses the zero-field splitting. But increasing the temperature also reduces the difference in population between spin-up and spin-down electrons. For $T \gg g\beta H_0$, this latter process exhibits a T^{-1} dependence. The actual reduction in EPR intensity is thus a convolution of the exponential and inverse processes, which leads to a precipitous decline in the signal amplitude (Figure 10).

The same figure also illustrates the second consideration, viz. aspects of the procedure required to obtain accurate intensity measurements of this species. There are two difficulties involved here.

First, one has to implement the mechanics of the integration procedure, a nontrivial consideration because about 50% of the EPR intensity lies between $g = 5$ and $g = 2$ (Figure 5). Aasa et al. [22] have established empirically a quantitative technique that relies only on measuring the intensity of the intense $g = 6$ feature.

Secondly, it is clearly important to know the fraction of molecules in the $m_s = \pm \frac{1}{2}$ state; thus the value of D (and E) needs to be established. This is not straightforward and requires evaluations of the EPR intensity at a number of temperatures and comparison with a standard for which D is known accurately [51].

D. MIXED-SPIN STATES

The phenomenon of mixed-spin states has several implications for hemoproteins. The most obvious example occurs in those situations where spin transitions are induced by an exogenous ligand but the amount of ligand present in solution is insufficient to saturate the heme. Although a conceptually trivial circumstance, this proves in practice to be quite important because the concentration of ligand needed to obtain full occupancy of the heme may be very high, approaching 1 M in unfavorable cases. Such high concentrations may not be attainable with less soluble compounds.

A more interesting example of mixed-spin behavior arises when the ligand field strength of the axial ligands is almost adequate to induce a high-spin to low-spin transition and the relative contributions of the two components can be modulated by temperature, with a reduction in temperature favoring the low-spin component. A classic example of this behavior is provided by metmyoglobin. With this protein the sixth ligand is water at low pH and hydroxide at high pH. The low-pH material is essentially high-spin, while at high pH the protein has about one-third low-spin character. Thus, on raising the pH there is a high-spin to low-spin conversion with a nominal pK of about 9. The proportions of the two spin states are temperature-dependent, and it is possible to define an equilibrium constant

$$K = a/(1-a), \qquad (17)$$

where a is the fraction of low-spin species. The variation of K with temperature yields the enthalpy and entropy for the equilibrium process [56].

The possibility of temperature-dependent spin-state variations underscores a classic hazard of the EPR technique. The data are almost always necessarily recorded at low temperature, but frequently the information obtained from the spectrum is correlated with additional data obtained at room temperature. The spin composition of the sample may well be different at the two temperatures. There is thus a need for spectroscopic techniques sensitive to the spin state that can be applied at room temperature and that can provide a standard of comparison between the data obtained at the two temperatures. Magnetic circular dichroism seems to be the most promising of the available methods.

The common characteristic of two categories of mixed-spin behavior just described is that the sample under study exhibits a spectrum that is the weighted sum of the two component spin systems. This is in marked contrast to the behavior of quantum-mechanically mixed spin systems, in which one obtains a spectrum that is the average of the participating species. The best example of this behavior is to be found in an unusual protein, cytochrome c' from Chromatium. In this protein the heme appears to be 5-coordinated with histidine as the single axial ligand [57].

The combination of a relatively weak axially directed field with a strong in-plane ligand field raise the energy of the $d_{x^2-y^2}$ orbital so that the ^4A intermediate-spin configuration of $(xy)^2(xz)^1(yz)^1(z^2)^1$ becomes comparable in stability to the usual ^6A high-spin state (Figure 9). This intermediate spin state is also split by spin-orbit processes, which lead to a stabilization of the $\pm \frac{1}{2}$ doublet relative to the $\pm \frac{3}{2}$ levels. The g-values of the lowest doublet are $g_\perp = 4$ and $g_\parallel = 2$ (Figure 9).

Mixed-Spin States

The anisotropy in the ligand fields are rarely so severe as to establish this intermediate-spin state as a well-isolated ground state, and that certainly does not seem to have happened in hemoproteins. Rather, it appears that the axial field is only adequate to render the ^4A and ^6A states of comparable stability, (Figure 9), whereupon the two configurations are mixed by spin-orbit coupling and the resultant ground state has the form

$$|\Psi\rangle = a\,|^6A_1\rangle + b\,|^4A\rangle, \tag{18}$$

with $a^2 + b^2 = 1$ and a/b increasing with increasing stability of the high-spin component. The g-values exhibited by this quantum-mechanically mixed state are

$$g_\perp = 6a^2 + 4b^2,$$
$$g_\| = 2. \tag{19}$$

In the Chromatium protein $g_\perp$ was found to be 4.75, implying that the ground state is composed of approximately 30% ^6A and 70% ^4A state. The magnetic moment predicted from the relationship $u^2 = 35a^2 + 15b^2$ is 24, compared with 26.5 measured experimentally at room temperature.

Originally, the general acceptance of this quantum-mechanical spin mixing was compromised by data from MCD and Mössbauer experiments, which suggested that the experimental observations on c' might be explained by inhomogeneities in the protein samples studied. However, the original interpretations [58] have now been put on much more solid ground by several reports of heme model compounds which exhibit the mixed-spin characteristic. Thus, Reed et al. [59] described the synthesis of perchlorato (meso-tetraphenyl porphinato) Fe(III) and established the heme to be five-coordinated with the Fe atom displaced about 0.3 Å towards the oxygen atom of the coordinating perchlorate. The observed magnetic moment of 4.5–5.3 Bohr magnetons is clearly too small for a $S = \frac{5}{2}$ paramagnet, while the Mössbauer spectrum [60] exhibits a quadrupole splitting much too large to be due to the essentially symmetric ^6A ground state. The EPR of the solid is axially symmetric with $g_\perp = 4.75$ and $g_\| = 2.03$. Each of these observations is consistent with the conclusion that the iron has substantial $S = \frac{3}{2}$ character in the solid. On dissolution in toluene the EPR spectrum changes and exhibits both high- and mixed-spin components; solutions in tetrahydrofuran are exclusively high-spin.

A similar sensitivity to solvents is observed in the perchlorate derivative of octaethylporphinato Fe(III) prepared by Dolphin and his colleagues [61] and deduced to be intermediate-spin from Mössbauer and EPR measurements. Solutions of OEP-Fe(III)ClO$_4$ in dichloromethane appear to maintain their intermediate-spin character, at least as judged by Raman and NMR measurements [62], but increasing the solvent polarity (e.g., by the addition of dimethyl sulfoxide) leads to the gradual conversion to the high-spin state.

The frequencies of the anomalously polarized core-size marker of OEP-Fe(III)ClO$_4$ in CH$_2$Cl$_2$ is 1558 cm^{-1}, close to that for low-spin compounds consistent with a vacancy in $d_{x^2-y^2}$; addition of DMSO lowers the frequency by 25 cm^{-1} to a value typical of 6-coordinated high-spin heme [62]. The core-size marker of cytochrome c' is at 1578 cm^{-1}, closer to the low-spin limit than to the 6-coordinated $S = \frac{5}{2}$ value, although this value is only slightly higher than the 1572 cm^{-1} observed in 5-coordinated hemes [63].

Detailed comparison of the properties of the protein with the model compounds is complicated by its marked sensitivity to pH, the hemeprotein being reversibly converted to the high-spin configuration at both pH extremes. Nevertheless, the susceptibility, EPR, Mössbauer, and Raman data obtained on this protein at neutral pH are most satisfactorily explained by the original hypothesis of Moss and Maltempo, who concluded that this protein is a quantum-mechanical mixture of about one-third high-spin and two-thirds intermediate-spin states.

E. THE EPR OF COMPOUND I OF PEROXIDASE

The peroxidase family of enzymes exhibit a catalytic cycle that can be represented [64]

$$[\text{Fe(III)}]^+ + \text{H}_2\text{O}_2 \rightarrow [\text{Fe(IV)}]^{3+} + 2\text{H}_2\text{O},$$
$$[\text{Fe(IV)}]^{3+} + \text{RH} \rightarrow [\text{Fe(IV)}]^{2+} + \text{R}^+, \qquad (20)$$
$$[\text{Fe(IV)}]^{2+} + \text{RH} \rightarrow [\text{Fe(IV)}]^+ + \text{R}^+.$$

In this scheme the resting enzyme containing ferric heme reacts with hydrogen peroxide and undergoes a formal two-electron redox reaction, with the peroxide being converted to water and the heme oxidized by two equivalents. This latter species, called compound I, is variously represented as Fe^{5+} or as Fe^{4+}P$^+$ (P$^+$ being the porphyrin cation radical), with the latter description currently finding more favor. Subsequently the enzyme is restored to its original oxidation level via compound II, by two consecutive one-electron reduction reactions, each occurring with the concomitant oxidation of an organic donor to its cation radical.

Compound I contains an odd number of electrons and thus is expected to be detectable by EPR. However, until recently all attempts to demonstrate the EPR signal of this species have been rather unconvincing. For example, Aasa et al. [65] observed that the addition of H$_2$O$_2$ to horseradish peroxidase led to the appearance of a narrow EPR signal at $g = 1.995$ which was maximally developed at a molar ratio of 1:1 H$_2$O$_2$:enzyme, but that the integrated intensity of this signal accounted for only 0.01 spins per iron. This abysmally low yield of paramagnet made it difficult to associate this species with the active form of the enzyme, for corresponding optical data implied that there should have been essentially complete conversion of the enzyme to the intermediate state.

Recognizing that the published spectra appeared to have a broad underlying component, Schultz et al. [66] recently undertook a reexamination of the EPR properties of this system with the specific aim of locating any broad resonances that might be present in the sample. As an experimental aid they adopted rapid-passage conditions.

By working at suitably low temperatures (< 4 K) it is possible to sweep the magnetic field through the individual micropopulations (spin packets) present in the sample in a time short compared to the electron-spin relaxation time. Then, by adjusting the spectrometer to respond to the dispersive component of the EPR signal, one can measure the EPR absorption envelope (Figure 2) directly. Provided one can satisfy a number of experimental constraints (see Weger [67]), this procedure provides a substantial increase in sensitivity for broad signals. Using this approach, Schultz et al. discovered that the narrow

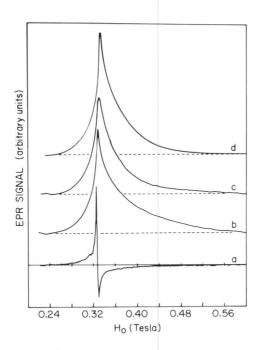

Figure 11 EPR spectra of compound I of horseradish peroxidase: (a) conventional (slow-passage) derivative spectrum; (b) numerical integration of (a); (c) rapid-passage absorption spectrum; (d) spectrum computed as described in the text. After Schultz et al. [66].

signal reported previously by Aasa et al. was simply the sharpest feature in an exceptionally broad and bland absorption envelope which extended over 3000 gauss (Figure 11) [66].

The measurement of the intensity of this species was complicated because conventional EPR spectra are needed to obtain accurate intensity data and much of the spectrum lies close to the baseline (Figure 11). However, with careful attention to sample preparation and to baseline correction procedures it was found that this EPR resonance could account for more than 0.8 electrons per heme [68], and this number was interpreted to mean that the primary intermediate of peroxidase contains one unpaired electron per heme.

In providing an explanation for this phenomenon, Schultz [69] drew on the results of Mössbauer measurements on these two intermediate forms of peroxidase. The Mössbauer data on compound II had been satisfactorily described in terms of a low-spin Fe(IV) center, with an electronic configuration of t_{2g}^4 and a net spin of 1. Because it is a spin-even system, the EPR is not (readily) observable, but the g-values deduced from the Mössbauer analysis were 2.25, 2.25, and 1.98, consistent with an inverted d^1 configuration. The zero-field splitting parameter D was calculated to be about 23 cm^{-1}, so that in zero field the spin levels are as shown in Figure 12, with the nonmagnetic singlet lying lowest.

Although compound I is one oxidizing equivalent above Compound II, the Mössbauer spectra of the two derivatives exhibit (1) the same isomer shift, (2) the same small, positive quadrupole splitting, and (3) absence of broad magnetic splitting in a weak applied magnetic field. These similarities imply that the valence electrons of the iron are unchanged in the transition from II to I and the extra oxidizing equivalent is assumed to reside elsewhere, most plausibly on the porphyrin ring, which would then be represented as a π-cation radical.

A strong antiferromagnetic interaction between these two paramagnetic centers can be eliminated, because the ground state would have a net spin of $\frac{1}{2}$ with g-values

$$g_x = g_y = \tfrac{4}{3}(2.25) - \tfrac{1}{3}(2.0) = 2.33,$$
$$g_z = \tfrac{4}{3}(1.98) - \tfrac{1}{3}(2.0) = 1.98. \tag{21}$$

The observed EPR clearly does not fit this description, while the Mössbauer spectra contradict the presence of a strong exchange coupling. However, the Mössbauer spectra of compound I do suggest the presence of a weak exchange interaction, of magnitude 2 cm^{-1}. This value is an order of magnitude less than the zero-field splitting present in compound II. Schultz and his colleagues explored this possibility and deduced that a weak exchange interaction would substantially explain the observed EPR spectrum. The coupling scheme is shown in Figure 12, which reveals that the nonmagnetic ground state ($S = 1$,

The EPR of Compound I of Peroxidase

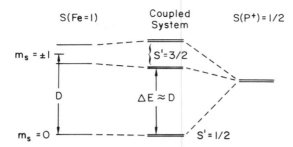

Figure 12 Energy-level scheme for compound I of horseradish peroxidase. Weak antiferromagnetic coupling between $S = 1$ (Fe^{+4}) and $S = \frac{1}{2}$ (P^+) leads to a ground state with net spin $\frac{1}{2}$ and an excited state with net spin $\frac{3}{2}$. Note that the antiferromagnetic interaction is much smaller than the zero-field splitting.

$m_S = 0$) of the iron is coupled to the porphyrin cation radical to produce a state with an effective spin of $\frac{1}{2}$.

Application of perturbation theory to this system gave

$$g'_x = 2.0 + g_x \frac{J_x}{D},$$
$$g'_y = 2.0 + g_y \frac{J_y}{D}, \quad (22)$$

where the unprimed g-values are those of the iron center and J refers to components of the exchange interaction. Because Mössbauer data are best fitted with J_x and J_y having the same sign, the extremely broad envelope of the EPR spectrum requires that there be "J-strain", a distribution in the magnitudes of J sufficiently broad to include both positive and negative values. With this assumption and the EPR parameters of Fe from compound II, it was possible to simulate the EPR spectrum using J-values of -2 (x), -1 (y) and 3 (z). (See Figure 11.)

Such exchange interaction is surprisingly small and implies the absence of any significant contact between the metal and ligand spins. The two metal electrons are in the $d\pi$ system of d_{xz} and d_{yz}, while the porphyrin "hole" is located in either the a_{1u} or the a_{2u} component of the near-degenerate pair of highest occupied molecular orbitals. The former orbital follows the outer margin of the porphyrin skeleton (Figure 13) and can be crudely approximated as a circle with a diameter of 8 Å [69]. The a_{2u}-orbital follows the inner margin of porphyrin macrocycle and approximate a circle of 4-Å diameter. A calculation [69] of the magnitude of the dipolar coupling between the two paramagnets yields values of -2, -2, and 4 cm^{-1} for an a_{2u}-radical and -0.3, -0.3,

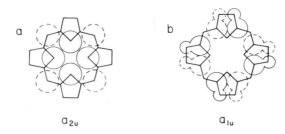

Figure 13 Representation of the two highest occupied symmetry orbitals of porphyrin.

and 0.6 cm^{-1} for an a_{1u}-radical. The former values are close to those used to simulate the EPR spectrum, consistent with the idea that the exchange interaction arises through dipolar coupling of the $S = 1$ Fe center and a porphyrin cation radical of a_{2u} symmetry.

The analysis predicts that a compound I in which the porphyrin radical is of a_{1u} symmetry would experience a much smaller dipolar coupling and exhibit a substantially modified EPR spectrum. This appears to be the case [68]. Horseradish peroxidase in which the protoheme is replaced by deuteroheme can also be converted to compound I, but in this instance the optical spectrum suggests that the porphyrin radical is of a_{1u} symmetry.* The EPR spectrum of this species shows a much more intense narrow component in the spectrum, being some 25-fold more intense than the corresponding signal in the protoheme protein. This increase in free-radical intensity is to be expected from the large decrease in magnetic-dipole interaction calculated for a spin present in the a_{1u}-orbital.

Further evidence in support of this formulation is provided by ENDOR. The ENDOR spectrum of horseradish peroxidase compound I exhibits features reasonably ascribed to the interaction of protons and nitrogen(s) with the unpaired electron. The same spectrum is obtained when the measurements are made at the center of the EPR resonance and in the wings (Figure 10), thus confirming that the EPR is associated with a single paramagnet. Furthermore, when compound I is prepared using heme deuterated at the β-carbons of the porphyrin ring, specific features in the ENDOR spectrum disappear [70]. This observation establishes unequivocally that the porphyrin ring is a component of the paramagnetic system and that the description of compound I as an antiferromagnet composed of Fe(IV) and porphyrin π-cation radical is on solid ground.

---

* This is established by comparison with electrochemically generated porphyrin cation radicals of presumed symmetry.

F. EPR OF NITROSYL HEMOPROTEINS

A characteristic property of high-spin ferrous hemoproteins is their ability to react with externally added, nonprotein ligands. Often these ligands have physiological significance in the functioning of the protein; the binding of oxygen by hemoglobin and by cytochrome oxidase are familiar and impressive examples. However, it is also possible to react the iron with reagents chosen to facilitate the experimental investigation of the hemeprotein system. The best example of such a reagent is carbon monoxide, for not only does this species bind avidly and rapidly to most high-spin ferrous hemes, but the resulting adduct is strongly photosensitive, and this property has been exploited in a variety of elegant and cunning scenarios [71, 72].

In the same vein, the reagent nitric oxide has found its place as a valuable probe of heme structure and function. This compound also reacts with high-spin ferrous heme rapidly and with high affinity according to the equation

$$Fe^{2+} \text{ (h.s.)} + NO(S = \tfrac{1}{2}) = Fe^{2+} \text{ (l.s.)}:NO(S = \tfrac{1}{2}). \qquad (23)$$

The electronic configuration of the reaction product can be represented simply as low-spin $3d^7(S = \tfrac{1}{2})$, the paramagnetic metal acquiring an additional electron from the NO ligand. However, an accurate description of the system is more complex (see below).

The salient property which leads to the experimental utility of this probe is the EPR characteristics of the heme-NO adduct. This spectrum is centered at $g = 2.00$ and is anisotropic, with both axial and rhombic species identifiable. However, the g-anisotropy is quite small, the extreme values of g_x and g_y being 2.08 and 1.95 respectively; g_z is relatively constant at 2.005.

Frequently, hyperfine structure is observed at g_z. In most cases nine lines can be resolved [Figure 14(c)]. From a comparison of the spectra obtained with ^{14}NO and ^{15}NO and from model systems of heme, NO, and nitrogenous bases, it is clear that the nine lines comprise a triplet of triplets due to an interaction between the paramagnetic electron and two nitrogen ($I = 1$) centers. The nitrogen of the nitrosyl function interacts strongly with the electron to yield a major splitting of about 21 gauss, while the nitrogenous base *trans* to the NO interacts only modestly and results in each component of the major triplet being split into three, the separation of the components of the minor triplet being about 7 gauss. When ^{15}NO ($I = \tfrac{1}{2}$) is used, the major triplet is replaced by a doublet and the major splitting increases to about 30 gauss, reflecting the 1.4-fold increase in magnetogyric ratio provided by ^{15}N.

Following the original characterization of these heme-nitrosyl species by Kon and Kataoka [73], Yonetani and his colleagues [74] extended the approach to a variety of hemoproteins and established in more detail the consequences of changing the nuclear spin either of the ligating nitrogen atoms or of the heme iron; these latter substitutions established that the unpaired electron also

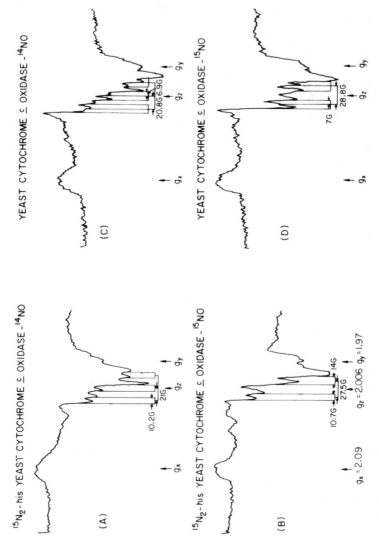

Figure 14 EPR spectra of nitrosyl cytochrome oxidase in which the NO and/or proximal histidine have been substituted with ^{14}NO or ^{15}N as shown. (a) ^{15}N-histidine, ^{14}NO; (b) ^{15}N-histidine, ^{15}NO; (c) ^{14}N-histidine, ^{14}NO; (d) ^{14}N-histidine, ^{15}NO. Modified from Stevens and Chan [75].

interacted with the central metal ion. Yonetani noted that g_z was essentially invariant, that g_x varied only slightly (2.05–2.08), and that g_y exhibited the greatest variability (1.955–2.01); it became apparent that good resolution of the hyperfine-splitting pattern required spectra that were essentially rhombic, for as g_y approached g_z the overlap in absorption reduced the depth of the individual hyperfine lines. The proteins studied by Yonetani fell into two classes. Most exhibited the "classical" nine-line pattern we have just discussed. However, two of them, catalase* and lactoperoxidase, exhibited a three-line spectrum suggesting that the ligand *trans* to the NO is not nitrogen in these proteins. A similar result for P450 was discussed in detail earlier in the chapter.

The most striking application of this simple concept is provided by recent results on cytochrome oxidase [75]. The oxygen-binding site of this enzyme (called cytochrome a_3) reacts avidly with NO, and the product of the reaction exhibits the typical nine-line EPR hyperfine spectrum; this spectrum changes appropriately when ^{15}NO is used. The most obvious conclusion to be drawn from this result is that histidine is the ligand *trans* to the NO in this heme center. However, as will be readily appreciated, there is a rather substantial logical jump from demonstrating the narrow triplet in the EPR spectrum to identifying the origin of that triplet with the imidazole moiety. Recognizing this limitation, Stevens and Chan [75] undertook the preparation of cytochrome oxidase in which the histidine residues contained ^{15}N. This was achieved by first creating a mutant of baker's yeast which has an obligatory requirement for histidine in its diet (a histidine auxotroph). This mutant was then cultured in media containing either ^{14}N-histidine or ^{15}N-histidine plus other necessary nutrients. The yeast was harvested and submitochondrial particles prepared from the cells. These particles were reacted with NO, and the EPR spectra recorded. The data are shown in Figure 14(a)–(d), which presents the four combinations of ^{14}N and ^{15}N histidine with ^{14}NO and ^{15}NO. The spectra exhibit the expected splitting patterns, ranging from a triplet of triplets for all ^{14}N to a doublet of doublets for all ^{15}N, with the changes in splitting constants reflecting the changes in nuclear moment in the required way. This very elegant result establishes clearly that histidine is the proximal ligand to the heme in cytochrome a_3, a conclusion of major relevance to efforts aimed at characterizing the structural attributes of this heme center.

The observation that g_z is essentially the free-electron value and that the *trans* ligand contributes significantly to the hyperfine splitting led to the early suggestion that the unpaired electron was present in the d_{z^2} orbital (an electron in d_{z^2} has no orbital angular momentum about the z-axis) and that the electronic configuration might be written simply as $3d^7$. However, this configuration requires all g-values to be greater than 2.0, and consequently fails to explain the fundamental result that g_y is typically less than 2.0. The very large

---

* In the case of beef liver catalase it is now established that the proximal heme ligand is tyrosine (Reid, T. J., Mathus, R. N., Sicignano, A., Tanaka, N. Musick, W. D. L. and Rossman, M. G. *Proc. Nat'l. Acad. Sci. USA* **78** (1981) 4767–4771.

hyperfine coupling with the nitrosyl group implies that the unpaired electron is, in fact, substantially confined to this ligand (compare the value of 21 gauss with the value of 32 gauss for A_{zz} found in nitroxide spin labels). In free NO the unpaired electron is present in a degenerate pair of $p\pi^*$ antibonding orbitals having cylindrical symmetry with respect to the N—O axis. In linear M—NO systems the two orbitals interact with the metal $d\pi$ orbitals (d_{xz}, d_{yz}), leading to the energy level scheme shown in Figure 15 [76]. As the M—NO bond is bent (in a direction we shall define as x), the interaction between $p\pi_x$ and d_{xz} is split and is replaced by an interaction between $p\pi_x$ and d_{z^2}; the bonding combination is represented as (π^*(NO), d_{z^2}) (Figure 15) to emphasize the larger contribution of the ligand orbital. Decreasing the M—NO angle increases the stability of this orbital, but as some lower-lying orbitals are slightly destabilized by the distortion, maximum stability of the system occurs at an intermediate angle, about 120°. Thus, the paramagnetic electron is placed in a ligand-centered orbital of z^2 symmetry giving a large hyperfine splitting and a value for g_z close to 2.0; g_x will be greater than 2.0 by an amount dictated by the energy separation Δ_x, while g_y will be governed by a tradeoff between the two contributions specified by Δ_y^+ and Δ_y^-. These will be quite variable and depend upon the bond angle. Furthermore, because the spin-orbit coupling for Fe is so much larger than that of the ligand atoms, the extent of d-orbital contribution to the wave functions mixed by the spin-orbit interaction will modulate the magnitudes of the g-shifts.

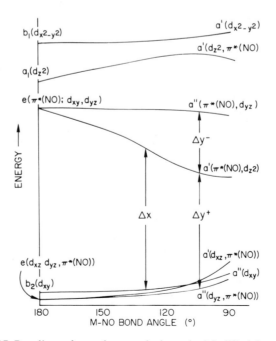

Figure 15 Bonding scheme for metal nitrosyls. Modified from [76].

Although the most frequent exploitation of this probe relies on the ability to resolve the nitrogen hyperfine contributions, there are a variety of other applications of the EPR spectra of these derivatives. This is most readily illustrated by reference to recent investigations on the properties of hemoglobin and its reaction with this ligand. First of all, it is known that the heme-NO adducts of the α- and β-chains of hemoglobin exhibit quite different EPR spectra, the α-chains exhibiting a spectrum that has essentially axial symmetry with weakly resolved structure on g_z, while the spectrum of the β-chains is almost isotropic with small inflections suggestive of underlying structure (Figure 16). It must be noted however that the spectra of both species are temperature-dependent. Most commonly, spectra are recorded at liquid-nitrogen temperatures and the differences just described are those that pertain at this high temperature. At 4.2 K, however, the spectral shape of the β-chains becomes asymmetric and resembles the spectrum of the α-chains recorded at 77 K (the spectra of the a-NO compound sharpens slightly on lowering the temperature to 4.2 K, although there is no profound change in overall shape). This phenomenon appears not to be due to any intrinsic differences in the α- and β-chains, for Morse and Chan [77] have shown that the EPR spectra of the myoglobin-NO complex contain the same two components, the proportion of the isotropic species increasing on raising the sample temperature. They labeled the anisotropic and isotropic species I and II respectively. (These species are not to be confused with two different derivatives studied by

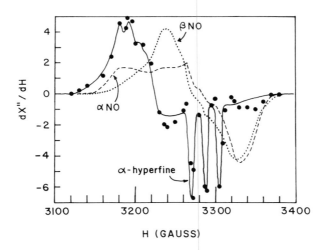

Figure 16 Standard EPR spectra of hexacoordinate a-NO and β-NO hemes and of pentacoordinated α-NO (designated α-hyperfine). The filled circles represent a spectrum obtained by subtracting the spectrum of β-NO heme from fully saturated Hb_4NO_4 containing bound IP_6 (the dashed line of Figure 17). The associated solid line is the experimental spectrum of 10% saturated hemoglobin incubated 10 minutes in the presence of IP_6 (see text). From [85].

Trittelvitz et al. [78], which were also labelled I and II and which will be discussed shortly.)

While there may be several explanations for this temperature dependence, the most plausible invokes mobility of the Fe—NO center in the β-chains such that there is the possibility of rotation about the Fe—N bond. With suitably rapid motion g_x and g_y would be averaged and an essentially isotropic spectrum would result. An alternative suggestion [77] requires a lengthening of the bond between the metal and the axial base (histidine) in species II. However, it is not clear why this should lead to a reduction in the hyperfine interaction (see below) or to the attenuation of the g-anisotropy. Whatever the explanation, it is clear that by choosing to work at suitably high temperatures it is possible to obtain significantly different contributions of the α-NO and β-NO chains, and this fact has been exploited by a number of investigators.

For example, it has been shown that the EPR spectrum of fully saturated Hb—NO is simply the spectrum of the separated α-NO and β-NO chains [79], that both hemes react at the same rate with NO when assembled in Hb [80], and that the EPR spectrum undergoes significant and often striking changes with change in pH or upon addition of organic phosphates [81].

In the absence of organic phosphates the EPR spectrum of Hb—NO does not show any clearly resolved ^{14}N hyperfine splittings. However, in 1972, Rein et al. [81] reported that addition of organic phosphates, e.g. inositol hexaphosphate (IP_6), to Hb—NO produced a marked change in the spectrum, a pronounced hyperfine triplet appearing at $g = 2.00$ (Figure 17) with a separation between extrema of about 16.5 gauss.* A similar splitting was observed in the model compound tetraphenylporphin [Fe(II):NO(TPP—NO)] [82], which raised the possibility that a hyperfine splitting of this magnitude was diagnostic of five-coordinated heme. Addition of a nitrogenous base to this system converted the EPR to the familiar nine-line spectrum typical of six-coordinated heme with two nitrogenous axial ligands. Additional support for this point of view has been provided by both infrared and resonance Raman measurements, which document the appearance of vibrational modes when IP_6 is added to Hb_4NO_4 [83]. However, as the overall lineshapes of the three-line spectra obtained with Hb and the model compound were quite different, the possibility that the bond between the iron and the proximal histidine could be broken was not widely accepted. A competing explanation was offered by Chevion et al. [84], who proposed that the three-line species was still six-coordinated but that N_1 of the proximal histidine is protonated, lowering the basicity of N_3; thus Chevion et al. pictured the three-line to nine-line transition as reflecting the deprotonation of N_1.

---

* Hemoglobin exists in two major conformational states. The $T-$ state (deoxyhemoglobin) is the low-affinity conformation and is present during the formation of Hb_4—NO. The R-state is the high-affinity conformation typified by oxyhemoglobin: It is present in the formation of Hb_4NO_4 from Hb_4NO_3. Organic phosphates such as IP_6 force an $R \rightarrow T$ transition.

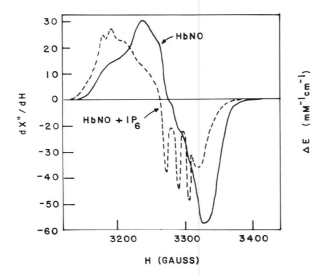

Figure 17 EPR Spectral changes associated with treating Hb_4NO_4 with saturating quantities of IP_6. From [85].

Hille et al. [85] undertook the systematic characterization of the EPR species present in mixtures of Hb and NO and of the effect of IP_6 on these mixtures. In this study they were cognizant of several facts. First, as already mentioned, the EPR of Hb_4NO_4 is accurately described by the 1 : 1 sum of the spectra of the nitrosyl compounds of the α- and β-chains. Second, although the combination of NO with Hb is very fast, there are time-dependent changes in the EPR spectrum that proceed over several minutes. During this time the intensity of the triplet increases markedly. Furthermore, the slowest phase appeared to represent the migration of NO from β-subunits to α-subunits (although the two kinds of chains react with NO at an equal rate, α-chains bind NO much more avidly than do β-chains). Hille et al. [85] observed that the contribution of the three-line species was maximal at low levels of saturation ($=Hb_4NO$) and decreased as the amount of NO bound increased. At all levels of saturation the addition of IP_6 converted the spectrum to one with maximal three-line character.

Anticipating that this new spectral species was due to a modification at the α-heme, they reasoned that this species would be maximally developed at low levels of saturation at long times in the presence of IP_6, conditions which would favor exclusive formation of the presumptive five-coordinated α-NO heme. Using the EPR spectrum that they recorded under these conditions (which they called a-hyperfine) together with the EPR spectra of isolated α- and β-chains (Figure 16), they attempted to fit the spectra of the HbNO species obtained at different levels of saturation with NO. They found that all three reference spectra were necessary to fit most of the experimental data. Thus, at low levels of saturation, the relative proportions of $\beta : \alpha : \alpha$-hyperfine were $0.5 : 0.25 : 0.25$, but as saturation increased the relative contributions of the two α-chains changed reciprocally, so that at complete saturation the system could be described by equal amounts of α- and β-chains with no contribution from the α-hyperfine species. In the presence of IP_6 very little of the normal α-species could be detected in the reaction mixture except when the system was fully saturated with respect to NO. It became clear that the α-chains in nitrosyl hemoglobin can exist in two conformations and that the relative proportions of these two conformations can be manipulated.

The EPR spectrum of the pure α-hyperfine species obtained by Hille et al. [85] is strikingly similar to the TPP—NO model compound prepared by Wayland and Olson [82], both in overall appearance and in the presence of the prominent triplet at g_z, and this similarity is persuasive evidence for the essential identity of these two species. Additional support for this conclusion is provided by the observation that flash photolysis of IP6-treated Hb—NO at 20 K leads to the formation of a species with the optical properties of a four-coordinated heme [86]. Furthermore, the competing point of view advocated by Chevion et al. [84] is weakened by recent resonance Raman measurements which show that the changes in the core-size marker that occur upon addition of IP_6 to Hb—NO are in the opposite direction to those obtained on proceeding from imidazole to imidazolate ligation [87]. Moreover, calculations suggest [21] that the magnitude of the hyperfine splitting associated with the minor hyperfine splitting is relatively insensitive to bond length and that the Fe—histidine bond must be increased by more than 1 Å to obtain the three-line pattern. In effect, one must break the bond. This theoretical result is supported by the report that the resonance Raman spectrum of peroxidase shows an unusually high frequency for the Fe—histidine stretch, which, from comparison with model compounds, appears to be due to the presence of the histidine as an imidazolate. But the EPR of the nitrosyl peroxidase (II) species exhibits the usual nine-line hyperfine pattern, implying that the EPR is indeed insensitive to the state of ionization of the proximal histidine.

This discussion has highlighted the rich variety of information available using NO as a heme probe. Simple lineshape analysis can be used to establish if a trans ligand exists and, if so, whether or not it is a nitrogen atom. Furthermore, it seems that the temperature of the spectrum which is observed in some systems might serve as a probe of the degree of steric hindrance present in the vicinity of the domain of the sixth coordination site. Such information is of direct relevance to dynamics of hemoprotein-ligand interaction, a research area much in vogue at this time [88, 89].

Acknowledgments

I would like to express my thanks to Dr. Peter Debrunner for reprints and preprints of his work on compound I of peroxidase, to Dr. Kurt Wuthrich for providing Figure 8, to Dr. N. R. Orme Johnson for Figure 11, to Dr. S. I. Chan for Figure 14, to Dr. S. P. J. Albracht for a preprint of Reference [49], and to Dr. K. Carter for his help in preparing the figures.

The preparation of this article was supported by grants from the NIH (GM 21337 and the Welch Foundation (C 636).

References

1. Mulks, C. F., Scholes, C. P., Dickinson, L. R., and Lapidot, A. *J. Am. Chem. Soc.* **101** (1979) 1645–1654 and references therein.
2. Palmer, G. In *The Porphyrins*, D. Dolphin (ed.), Vol. IV, Academic, New York, 1979, pp. 313–353.
3. Blumberg, W. E. In *Magnetic Resonance in Biological Systems*, A. Ehrenberg, B. G. Malmstrom, and T. Vanngard (eds.), Pergamon, Oxford, 1967, pp. 119–135.
4. Griffith, J. S. *Nature (London)* **180** (1957) 30–31.
5. Kotani, M. *Prog. Theoret Phys. Suppl.* **17** (1961) 4–13.
6. Weissbluth, M. In *Hemoglobin: Cooperativity and Electronic Processes*, Springer, Berlin and New York, 1973.
7. Taylor, C. P. S. *Biochim. Biophys. Acta* **491** (1977) 137–149.
8. Bohan, T. H. *J. Magn. Res.* **26** (1977) 109–118.
9. Blumberg, W. E., and Peisach, J. In *Probes of Structure and Function of Macromolecules and Membranes*, B. Chance, T. Yonetani, and A. S. Mildvan (eds.), Academic, New York, 1971, Vol. 2, pp. 215–229.
10. Huynh, B. H., Emptage, M. H., and Munck, E. *Biochim. Biophys. Acta* **534** (1978) 295–306.
11. Chevion, M., Peisach, J., and Blumberg, W. E. *J. Biol. Chem.* **252** (1977) 3637–3645.
12. Champion, P. M., and Gunsalus, I. C. *J. Am. Chem. Soc.* **99** (1977) 2000–2002.
13. Teraoka, J., and Kitagawa, T. *J. Biol. Chem.* **256** (1981) 3969–3977.
14. Tang, S. P. W., Spiro, T. G., Mukai, K., and Kimura, T. *Biochem. Biophys. Res. Comm.* **53** (1973) 869–874.
15. Kitagawa, T., Abe, M., Kyoguku, Y., Ogoshi, H., Watanabe, E., and Yoshida, Z. *J. Phys. Chem.* **80** (1976) 1181–1186.
16. Cramer, S. P., Dawson, J. H., Hodgson, K. O., and Hager, L. P. *J. Am. Chem. Soc.* **100** (1978) 7282–7290.
17. Hanson, L. K., Sligar, S. G., and Gunsalus, I. C. *Croatica Chemica Acta* **49** (1977) 237–250.
18. Griffin, B. W., and Peterson, J. A. *J. Biol. Chem.* **250** (1975) 6445–6451.
19. Lo Brutto, R., Scholes, C. P., Wagner, G. R., Gunsalus, I. C., and Debrunner, P. G. *J. Am. Chem. Soc.* **102** (1980) 1167–1169.
20. Ebel, R. E., O'Keeffe, D. H., and Peterson, J. A. *J. Biol. Chem.* **253** (1978) 3888–3897.
21. Mum, S. K., Chang, J. C., and Das, T. P. *Biophys. J.* **25** (1979) 127a.
22. Aasa, R., Albracht, S. P. J., Falk, K. E., Lanne, B., and Vanngard, T. *Biochim. Biophys. Acta* **422** (1976) 260–272.
23. Babcock, G. T., Van Steelandt, J., Palmer, G., Vickery, L., and Salmeen, I. In *Cytochrome Oxidase*, T. E. King et al. (eds.) Elsevier North Holland, New York, 1979, pp. 105–115.
24. Blasie, J. M., Erecinska, M., Samuels, S., and Leigh, J. S. *Biochim. Biophys. Acta* **501** (1978) 33–52.

25. Blum, H., Harmon, H. J., Leigh, J. S., Salerno, J. C., and Chance, B. *Biochim. Biophys. Acta* **502** (1978) 1–10.
26. Brautigan, D. L., Feinberg, B. A., Hoffman, B. M., Margoliash, E., Peisach, J., and Blumberg, W. E. *J. Biol. Chem.* **252** (1977) 574–582.
27. Lambeth, D., Campbell, K., Zand, R., and Palmer, G. *J. Biol. Chem.* **248** (1973) 8130–8136.
28. Carter, K. R., Tsai, A. L., and Palmer, G. *FEBS Letters* **132** (1981) 243–246.
29. Salmeen, I., and Palmer, G. *J. Chem. Phys.* **48** (1968) 2049–2052.
30. Senn, H., Keller, R. M., and Wuthrich, K. *Biochem. Biophys. Res. Comm.* **92** (1980) 1362–1369.
31. Keller, R. M., Schejter, A., and Wuthrich, K. *Biochim. Biophys. Acta* **626** (1980) 15–22.
32. Dwivedi, A., Toscano, W. A., Jr., and Debrunner, P. G. *Biochim. Biophys. Acta* **576** (1979) 502–508.
33. Senn, H., and Wuthrich, K. Results presented at the Symposium on Single Electron Transfer in Biology, Konstanz, Germany, 1981.
34. Keller, R., Picot, D., and Wuthrich, K. *Biochim. Biophys. Acta* **580** (1979) 259–265.
35. De Vries, S., Albracht, S. P. J., and Leeuwerik, F. J. *Biochim. Biophys. Acta* **546** (1979) 316–333.
36. Siedow, J. N., Power, S., de la Rosa, F. F., and Palmer, G. *J. Biol. Chem.* **253** (1978) 2392–2399.
37. Ikeda, M., Iizuka, T., Takao, H., and Hagihara, B. *Biochim. Biophys. Acta* **336** (1974) 15–24.
38. Siedow, J. N., Vickery, L. E., and Palmer, G. *Arch. Biochem. Biophys.* **203** (1980) 101–107.
39. Mims, W. B., and Peisach, J. *J. Chem. Phys.* **64** (1976) 1074–1091.
40. Keller, R., Groudinsky, O., and Wuthrich, K. *Biochim. Biophys. Acta* **427** (1976) 497–511.
41. Keller, R. M., and Wuthrich, K. *Biochim. Biophys. Acta* **533** (1978) 195–208.
42. Senn, H., Keller, R. M., and Wuthrich, K. *Biochem. Biophys. Res. Comm.* **92** (1980) 1362–1369.
43. Keller, R. M., and Wuthrich, K. *Biochem. Biophys. Res. Comm.* **83** (1978) 1132–1139.
44. Keller, R. M., and Wuthrich, W. *Biochim. Biophys. Acta* **621** (1980) 204–217.
45. Aasa, R., and Vanngard, T. *J. Magn. Res.* **17** (1975) 308–315.
46. Abragam, A., and Bleaney, B. *Electron Paramagnetic Resonance of Transition Ions*, Clarendon, Oxford, 1970, p. 100.
47. Babcock, G. T., Vickery, L. E., and Palmer, G. *J. Biol. Chem.* **251** (1976) 7907–7919.
48. Sands, R. H., and Dunham, W. R. *Quart. Rev. Biophys.* **7** (1974) 443–504.
49. Hagen *J. Magn. Res.* **44** (1981) 447–469.
50. Feher, G. In *Electron Paramagnetic Resonance with Applications to Selected Problems in Biology*, Gordon & Breach, New York, 1970.
51. Peisach, J., Blumberg, W. E., Lode, E. T., and Coon, M. J. *J. Biol. Chem.* **246** (1971) 5877.
52. Alpert, Y., Conder, Y., Tuckendler, J., and Thome, H. *Biochim. Biophys. Acta* **322** (1973) 34–37.

53. Kotani, M. *Ann. N.Y. Acad. Sci.* **158** (1973) 20–49.
54. Brackett, G., Richards, P., and Caughey, W. S. *J. Chem. Phys.* **56** (1971) 4383–4401.
55. Scholes, C. P. *J. Chem. Phys.* **52** (1970) 4890–4895.
56. Iizuka, T., and Yonetani, T. *Advan. Biophys.* **1** (1970) 155–178.
57. Weber, P. R., Bartsch, R. G., Cusanovich, M. A., Hamlin, R. C., Howard, A., Jordan, S. R., Kamen, M. D., Meyer, T. E., Weatherford, D. W., Xuong, Nguyen huu, and Salemme, F. R. *Nature* **286** (1980) 302–304.
58. Maltempo, M., Moss, T. H., and Cusanovich, M. *Quart. Rev. Biophys.* **9** (1970) 181–215.
59. Reed, C. A., Mashiko, T., Bentley, S. P., Kastner, M. E., Scheidt, W. R., Spartalian, K., and Lang, G. *J. Am. Chem. Soc.* **101** (1979) 2948–2958.
60. Spartalian, K., Lang, G., and Reed, C. A. *J. Chem. Phys.* **71** (1979) 1832–1837.
61. Dolphin, D. H., Sams, J. R., and Tsin, T. B. *Inorg. Chem.* **16** (1977) 711–713.
62. Ogoshi, H., Sugimoto, H., and Yoshida, Z. *Biochim. Biophys. Acta* **621** (1980) 19–28.
63. Woodruff, W. H., Kessler, R. J., Ferris, R. S., Dallinger, R. F., Carter, K. R., Artalis, T. A., and Palmer, G. In *Advances in Chemistry*, K. Kadish (ed.), in press, and references therein.
64. Hewson, W. D., and Hager, L. P. (1979) In *The Porphyrins*, D. Dolphin, (ed.), Academic, New York, Vol. 7, pp. 295–332.
65. Aasa, R., Vanngard, T., and Dunford, H. B. *Biochim. Biophys. Acta* **391** (1975) 259–264.
66. Schultz, C. E., Devaney, P. W., Winkler, H., Debrunner, P. G., Doan, N., Chiang, R., Rutter, R., and Hager, L. P. *FEBS Letters* **103** (1980) 102–105.
67. Weger, M. *Bell Syst. Tech. J.* **39** (1960) 1013–1112.
68. Rutter, R., Hager, L. P., and Debrunner, P. In *Symposium on Interaction between Iron and Proteins in Oxygen and Electron Transport*, C. Ho (ed.), 1982.
69. Schultz, C. E., Dissertation, University of Illinois, 1979.
70. Roberts, J. E., Hoffman, B. M., Rutter, R., and Hager, L. P. *J. Biol. Chem.* (1981).
71. Gibson, Q. H. *Prog. Biophys. and Biophys. Chem.* **9** (1960) 1–53.
72. Gibson, Q. H., Greenwood, C., Wharton, D. C., and Palmer, G. *J. Biol. Chem.* **240** (1965) 888–894.
73. Kon, J., and Kataoka, K. *Biochemistry* **8** (1969) 4757–4762.
74. Yonetani, T., Yamamoto, H., Erman, J. E., Leigh, J. S., and Reed, G. H. *J. Biol. Chem.* **247** (1972) 2447–2455.
75. Stevens, T. H., and Chan, S. I. *J. Biol. Chem.* **256** (1981) 1069–1071.
76. Mingos, D. M. P. *Inorg. Chem.* **12** (1973) 1209–1211.
77. Morse, R. H., and Chan, S. I. *J. Biol. Chem.* **255** (1980) 7876–7882.
78. Trittelitz, E., Gersonde, K., and Winterhalter, K. H. *Eur. J. Biochem.* **51** (1975) 33–42.
79. Reisberg, P., Olson, J. S., and Palmer, G. *J. Biol. Chem.* **251** (1976) 4379–4383.
80. Hille, R., Palmer, G., and Olson, J. S. *J. Biol. Chem.* **252** (1977) 403–405.
81. Rein, H., Ristau, O., and Scheler, W. *FEBS Letters* **24** (1978) 24–26.
82. Wayland, B. B., and Olson, L. W. (1974) *J. Am. Chem. Soc.* **96**, 6037–6041.
83. Scholler, D. M., Wang, M. Y. R., and Hoffman, B. M. *J. Biol. Chem.* **254** (1979) 4072–4078.

84. Chevion, M., Stern, A., Peisach, J., Blumberg, W. E., and Simon, S. *Biochemistry* **17** (1978) 1745–1750.
85. Hille, R., Olson, J. S., and Palmer, G. *J. Biol. Chem.* **254** (1979) 12110–12120.
86. Stong, J. D., Burke, J. M., Daly, P., Wright, P., and Spiro, T. G. *J. Am. Chem. Soc.* **102** (1980) 5815–5820.
87. Nagai, K., Hori, H. Yoshida, S., Sakamoto, H., and Morimoto, H. *Biochim. Biophys. Acta* **532** (1978) 17–28.
88. Alberding, N., Chan, S. S., Eisenstein, L., Gunsalus, I. C., Nordlund, T. M., Perutz, M. F., Reynolds, A. H., and Sorensen, L. B. *Biochemistry* **17** (1978) 43–51.
89. Reisberg, P., and Olson, J. S. *J. Biol. Chem.* **255**, 4159–4169 (1980).

3
The Resonance Raman Spectroscopy of Metalloporphyins and Heme Proteins

THOMAS G. SPIRO

ABBREVIATIONS

RR—resonance Raman, i.r.—infrared, TPP PP—protoporphyrin IX, OEP—octaethylporphyrin, MP—mesoporphyrin IX, EP—etioporphyrin, Pa—porphyrin a, TPP—*meso*-tetraphenylporphine, Hb—hemoglobin, Mb—myoglobin, cyt—cytochrome, ImH—imidazole, 2-MeImH—2-methylimidazole, 1,2-diMeIm—1,2-dimethylimidazole, IHP—inositol hexaphosphate, THF—tetrahydrofuran, DMSO—dimethylsulfoxide.

A. INTRODUCTION

The fiftieth anniversary of the discovery by C. V. Raman of discrete lines in the spectrum of light scattered from molecules was celebrated in 1978 [1, 2]. During the 1930s Raman spectroscopy provided basic information on molecular vibrations [3]. The main thrust of vibrational studies soon passed to infrared spectroscopy, however, as infrared detectors and optical elements were developed. Because of the weakness of the Raman effect, heroic measures were required to obtain Raman spectra, usually involving the insertion of a large sample inside a cylindrical high-intensity metal-vapor lamp. Few samples of chemical or biological interest could fulfill the resulting requirements of size, optical clarity, and transparency. These requirements were dramatically relaxed in the late 1960s with the introduction of laser light sources, which revolutionized the art of Raman spectroscopy [4]. Because the laser produces an intense directional beam of light, which can be focused to a point, samples of small size and a wide range of optical properties can be examined.

Particularly important is the ability to obtain spectra of absorbing molecules, by arranging the experiment to minimize the laser light path through the sample, e.g. via backscattering, or via 90° scattering from a capillary, or from the corner of a sample cell. This capability has made possible the systematic study of the resonance Raman effect, in which Raman bands associated with vibrational modes that lead to excited-state distortions are enhanced, as the laser frequency approaches the electronic transition frequency [5]. This effect is of much intrinsic interest, and novel experimental findings have stimulated a great deal of activity in the theory of resonance enhancement. In addition, the effect is of practical importance for the study of biological and other complex molecular systems [6-8]. Resonance enhancement permits examination of the vibrational modes of the chromophores in a sample, unobscured by the vibrational modes of the molecular matrix. It provides a means of selecting out specific chemical groups by tuning the laser to their electronic transitions.

Porphyrins are attractive targets for resonance Raman (RR) spectroscopy, because of their intense absorption bands in the visible and near ultraviolet regions. RR spectra were first reported for hemoglobin, a decade ago [9-11]. Since then porphyrin RR spectroscopy has burgeoned, because of its potential for structural elucidation of hemeproteins, and also because the special molecular and electronic structure of porphyrin provides a wealth of novel resonance effects in Raman spectra. There have been several reviews [12–17], and the pace of new results continues to quicken. At this point there is a reasonably clear picture of the basic resonance enhancement mechanisms, the nature of the porphyrin vibrational modes, and the systematics of vibrational frequency shifts seen in the various common states of hemeproteins and their analog porphyrin complexes.

B. RESONANCE ENHANCEMENT

i. Basic Concepts

The resonance Raman effect has attracted much attention from theorists as well as experimentalists. Numerous theoretical treatments have been given [18–23], and increasingly powerful approaches have been developed to account for the rich variety of spectroscopic phenomena that have been uncovered. In this section, the basic concepts of resonance Raman scattering are briefly described.

For a Raman transition between molecular states g (ground) and f (final), the scattered light intensity [18] is

$$I_s = \frac{8\pi \nu_s^4}{9c^4} I_0 \sum \left| (\alpha_{\rho\sigma})_{gf} \right|^2, \tag{1}$$

where I_0 is the incident intensity at frequency ν_0, ν_s is the scattering frequency, c is the velocity of light, and $(\alpha_{\rho\sigma})_{gf}$ is the transition polarizability tensor, with incident and scattered polarizations indicated by ρ and σ. Second-order perturbation theory gives the Kramers-Heisenberg equation for the polarizability:

$$(\alpha_{\rho\sigma})_{gf} = \frac{1}{\hbar} \sum_e \frac{\langle f | \mu_\rho | e \rangle \langle e | \mu_\sigma | g \rangle}{\nu_{eg} - \nu_0 + i\Gamma_e} + \frac{\langle f | \mu_\sigma | e \rangle \langle e | \mu_\rho | g \rangle}{\nu_{ef} + \nu_0 + i\Gamma_e}, \tag{2}$$

where μ_ρ and μ_ρ are dipole moment operators, $|g\rangle$ and $|f\rangle$ are the initial- and final-state wave functions, $|e\rangle$ is the wave function of an excited state, of half-width Γ_e, and ν_{eg} and ν_{ef} are transition frequencies. When $\nu_0 \ll \nu_{eg}$, the two terms give comparable contributions and the polarizability is nearly independent of wavelength. As ν_0 approaches ν_{eg}, the first term becomes dominant and is responsible for resonance effects.

The wave functions may be separated into electronic and vibrational parts via the Born-Oppenheimer approximation, giving

$$\langle f | \mu | e \rangle = \langle j | M_e | v \rangle \quad \text{and} \quad \langle e | \mu | g \rangle = \langle v | M_e | i \rangle,$$

where $|i\rangle$ and $|j\rangle$ are the initial and final vibrational wave functions of the ground electronic state, and $|v\rangle$ is a vibrational wave function of the excited electronic state e. M_e is the pure electronic transition moment between g and e. It is a weakly varying function of the nuclear coordinates, and may be expanded in a Taylor series

$$M_e = M_e^0 + \left(\frac{\partial M}{\partial Q}\right)^0 Q + \cdots, \tag{3}$$

where Q is a given normal mode of the molecule. The first two terms in the series give

$$\alpha = A + B, \tag{4}$$

$$A = (M_e^0)^2 \frac{1}{\hbar} \sum_v \frac{\langle j|v\rangle\langle v|i\rangle}{\nu_{vi} - \nu_0 + i\Gamma_v}, \tag{5}$$

$$B = M_e^0 \left(\frac{\partial M}{\partial Q}\right)^0 \frac{1}{\hbar} \sum_v \frac{\langle j|Q|v\rangle\langle v|i\rangle + \langle j|v\rangle\langle v|Q|i\rangle}{\nu_{vi} - \nu_0 + i\Gamma_v}, \tag{6}$$

neglecting the nonresonant part of Equation (2), and dropping the polarization subscripts.

Since A is the leading term, it is ordinarily the dominant contribution to the RR intensity. A-term enhancement varies directly with the strength of the electronic transition and inversely with its bandwidth, and it depends on the magnitude of the Frank-Condon products $\langle j|v\rangle\langle v|i\rangle$. These increase with the extent of displacement of the excited-state potential well along the normal coordinate (origin shift). Thus enhancement for a given mode depends on the extent to which it is involved in the excited-state distortion. Indeed, the geometry of the excited state can in principle be mapped out from an analysis of the relative intensities of the RR bands [13].

Only totally symmetric modes can be enhanced via the A-term, since for non-totally-symmetric modes the origin shift, and therefore the Frank-Condon product, is zero. (Since j and i must differ for Raman transitions, either $\langle j|v\rangle$ or $\langle v|i\rangle$ must be zero in the absence of an origin shift). Non-totally-symmetric modes can gain intensity via the B-term because of the Q-dependent vibrational integrals. The normal coordinate connects vibrational levels differing by one quantum, so that $\langle 1|Q|0\rangle\langle 0|0\rangle$ and $\langle 1|1\rangle\langle 1|Q|0\rangle$ both contribute to the B-term for Raman fundamentals. To evaluate the B-term transition-moment derivative $(\partial M_e/\partial Q)^0$, one can apply the Herzberg-Teller approach [18] and consider it to arise from the mixing of the resonant state, e, with a nearby excited state, s. The resulting expression for the B-term is

$$B = \frac{(M_s^0)(M_e^0)h_{es}}{\nu_s - \nu_e} \sum_v \frac{\langle j|Q|v\rangle\langle v|j\rangle + \langle j|v\rangle\langle v|Q|i\rangle}{\nu_{vi} - \nu_0 + i\Gamma_v}, \tag{7}$$

where $h_{es} = \langle s|\partial H/\partial Q|e\rangle$, the integral that mixes the two states via a given normal coordinate. The B-term becomes important in cases where a forbidden or weakly allowed transition gains intensity from vibronic mixing with a strongly allowed transition. The mixing modes are then prominent in the RR spectrum when excited at the weak transition. The enhancement depends on the size of the mixing integral and on the proximity of the electronic states. A limiting case is the Jahn-Teller effect, in which the mixing states are degenerate. The Herzberg-Teller perturbation approach no longer applies in this case, but the Jahn-Teller active modes are strongly enhanced in the RR spectrum.

ii. Porphyrin π-π* Transitions

The electronic spectra of hemeproteins are discussed in detail by Makinen and Churg [24] in Chapter 3 of Volume I. Only those features of importance to the available RR results are recapitulated here. Metalloporphyrin absorption spectra are dominated by a very strong band ($\varepsilon \sim 10^5$ M cm^{-1}) near 400 nm, called the Soret band or B-band. Near 550 and 500 nm are two weaker bands ($\varepsilon \sim 10^4$ M cm^{-1}), called α and β, or Q_0 and Q_1. This basic three-band pattern can be understood on the basis of Simpson's [25] π-electron model for $4n + 2$ electron annulenes, in which the first two π-π* transitions occur between the degenerate pair of highest filled molecular orbitals, with angular momentum $l = \pm n$, to the lowest unfilled pair, with $l = \pm(n + 1)$. The two (doubly degenerate) transitions have $\Delta l = 2n + 1$ and $\Delta l = 1$, in order of increasing energy. The $\Delta l = 1$ transition is allowed, while the $\Delta l = 2n + 1$ transition is forbidden; but the latter can gain intensity from the former by vibronic mixing. Porphyrin dianion ligands approximate 16-atom (the inner porphyrin ring, see Figure 1) annulenes, with $n = 4$. The B and Q_0 bands correspond to the $\Delta l = 1$ and $\Delta l = 9$ transitions, while the Q_1 band is the envelope of 0-1 vibronic transitions, involving mixing modes with $\simeq 1500$-cm^{-1} weighted average vibrational frequency.

The maximum symmetry of metalloporphyrins is actually D_{4h}, rather than D_{16h}, and the highest filled annulene orbital pair is split into a_{1u} and a_{2u} orbitals, which, however, remain close in energy. The lowest unfilled orbital pair, e_g, remains degenerate. The two promotions $a_{1u} \to e_g$ and $a_{2u} \to e_g$ both have E_u symmetry, and are subject to strong configuration interaction [26]. The transition dipoles add for the B-transition, and nearly cancel for the Q_0-transition, accounting for its low, but nonzero intensity.

a. B-Band Scattering. This four-orbital model [26] goes far toward explaining the main features of metalloporphyrin RR spectra [27]. These are dominated by bands with frequencies in the range 1100–1650 cm^{-1}, corresponding to stretching of the porphyrin-ring π-bonds, as expected for enhancement via π-π* transitions. As an illustration of the general appearance of these spectra, Figure 2 shows RR spectra of Fe^{II} cytochrome c obtained with 4131- and 5309-Å excitation; these wavelengths are close to resonance with the B and Q transitions (see Figure 3).

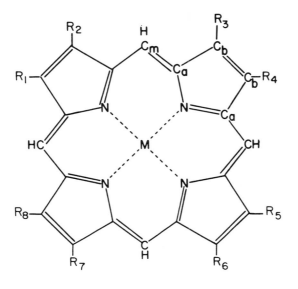

	R_1	R_2	R_3	R_4	R_5	R_6	R_7	R_8
Protoporphyrin IX	Me	V	Me	V	Me	P	P	Me
Deutero		H		H				
Meso		Et		Et				
Hemato		CHMe OH		CHMe OH				
Etioporphyrin I	Me	Et	Me	Et	Me	Et	Me	Et
Octaethylporphyrin	Et	Et	Et	Et	Et	Et	Et	Et

Me CH$_3$
Et CH$_2$CH$_3$
V CH=CH$_2$
P C$_2$H$_4$COOH

Figure 1 Structural diagram of the porphyrin ring with atom labeling. The substituent patterns are given for several physiological-type porphyrins. In tetraphenylporphine (TPP) the R groups are replaced by H, and the methine H atoms are replaced by phenyl rings.

Excitation near the intense B-band enhances totally symmetric modes, via A-term scattering [28]. The strongest feature is a band near 1360 cm^{-1}, corresponding to the breathing mode of the C—N bonds [29]. Even for this mode, the origin shift is small, as shown by analysis of the resonance CARS (coherent anti-Stokes Raman scattering) lineshapes [30], and as is also evident in the lack of any pronounced vibrational structure on the B absorption band [24]. The large size of the porphyrin π-system keeps the excited-state distortion relatively small. The impressive enhancements that are seen with excitation in the B-band [31] (satisfactory spectra can be obtained with solutions as dilute as 10^{-5} M) are due to the large electronic transition moment.

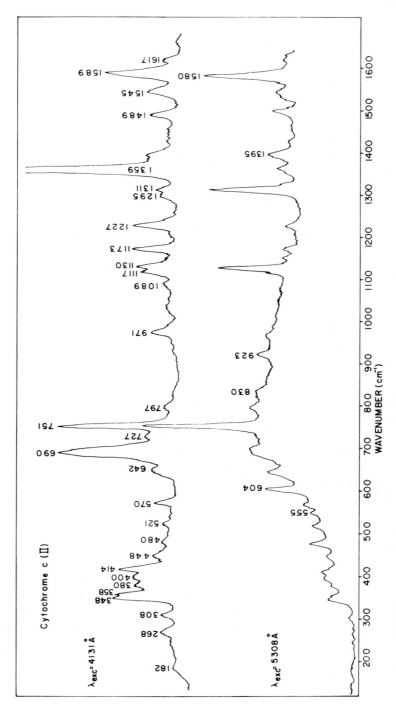

Figure 2 RR spectra of reduced cytochrome c (~1 mM) obtained with B (4131 Å) and Q (5309 Å) excitation.

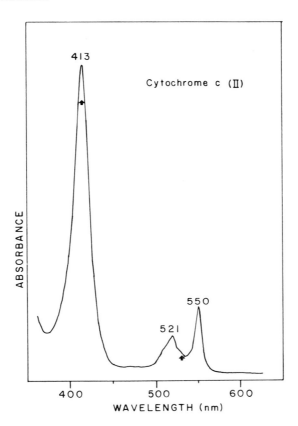

Figure 3 Absorption spectrum of reduced cytochrome c. The arrows mark the excitation wavelengths for the RR spectra shown in Figure 2.

For small origin shifts, the simple scattering theory outlined above predicts equal enhancement at the electronic origin (0-0) and at the position of the first vibronic level (0-1). This is because the first two terms of Equation (5) have Frank-Condon products, $\langle 1|0\rangle\langle 0|0\rangle$ and $\langle 1|1\rangle\langle 1|0\rangle$, of equal magnitude (but opposite sign) in the limit of small origin shifts. However, Champion and Albrecht [32] have shown that the excitation profile (plot of Raman intensity versus excitation wavelength) of the Fe^{II} cytochrome c 1362-cm^{-1} band is skewed to the blue side of the B absorption band (Figure 4). They account for this behavior in terms of the influence of the many low-frequency modes a molecule as complicated as porphyrin must possess; these produce a strong intensification of the excitation profile, but not of the absorption band, at the 0-1 energy. The scattering equations given above are written in the single-mode approximation, but the vibrational wave function of a polyatomic molecule contains all of its normal modes. As emphasized by Warshel and coworkers

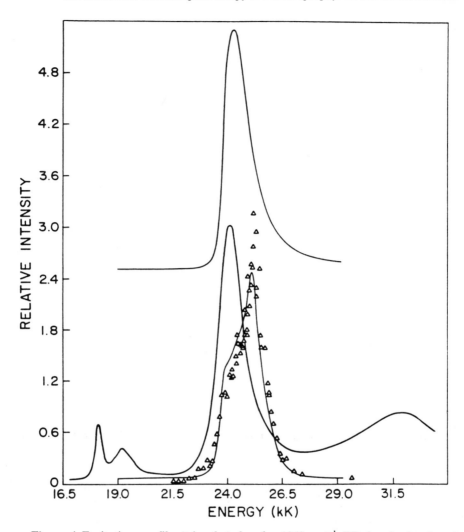

Figure 4 Excitation profile (triangles) for the 1362-cm^{-1} RR band of reduced cytochrome c, superimposed on the absorption spectrum. The profile peaks on the high-energy side of the B-band, at the position of the 0-1 transition. The thin lines are the calculated excitation profile and absorption band (shifted up for clarity) using the multimode theory developed in Reference [36] (Figure from [36]).

[33], there are Raman transition pathways in which modes other than the one being monitored are promoted in the resonant electronic state. If the Frank-Condon factor for a secondary, or "helping" mode [34] is large, then a subsidiary resonance can be observed. For example in both ClCrTPP [34] and O(FeTPP)$_2$ [35] (TPP = tetraphenylporphine), modes near 400 cm^{-1} were found to show excitation-profile maxima at the 0-1 (Q_1) position of the strong

≃ 1370-cm^{-1} band. Usually terms involving subsidiary modes are small, but they can affect the shape of the excitation profile, and the influence of many low-frequency subsidiary modes appears to explain the skewed profile observed by Champion and Albrecht [36]. (An alternative explanation involving interference between A- and B-term contributions to the scattering was advanced by Shelnutt [37a], but was shown by Champion and Albrecht [37b] to be incompatible with the observed differences between the widths of the Q and B bands.) Inclusion of these low-frequency modes also permits the use of narrow electronic bandwidths [36] in accounting for the excitation profile. It has been a conceptual difficulty of single-mode treatments that they often require large bandwidths, on the order of 1000 cm^{-1}, to account for the observed excitation profiles of polyatomic molecules, although electronic lifetimes are unlikely to be as short as such bandwidths imply [38].

b. Q-Band Scattering. With excitation in the Q_0 and Q_1 bands, A-term enhancement is much lower, because of the smaller electronic transition moment. The A-term scales with $(M_e^0)^2$, so that Q-band enhancement would be expected to be at least two orders of magnitude lower than B-band enhancement, if bandwidths and Frank-Condon factors were equal. The Q-band RR spectra are generally dominated by the non-totally-symmetric modes responsible for Q-B mixing and the generation of the Q_1-absorption. The symmetry of these modes is given by the direct product of the electronic transition symmetries, $E_u \times E_u = A_{1g} + A_{2g} + B_{1g} + B_{2g}$. Because of the high symmetry of the molecule, A_{1g} modes are ineffective in mixing the transition moments [39]. The remaining three symmetries are all represented among the modes observed in Q-band RR spectra. (B_{1g} and B_{2g} modes cannot be readily distinguished, since both give depolarized Raman bands, but they can be identified, via symmetry lowering, in the sharp-line fluorescence [11] or RR spectra [40] of free-base porphyrins, since $B_{1g} \to A_g$ but $B_{2g} \to B_{1g}$ when the D_{4h} symmetry is lowered to D_{2h}.)

Of particular interest are the A_{2g} modes, which give anomalously polarized RR bands [11]. For randomly oriented molecules, the depolarization ratio, which is the intensity ratio of scattered light with polarization perpendicular and parallel to the incident polarization, is [11, 5]

$$\rho = \frac{I_\perp}{I_\parallel} = \frac{3\gamma_s^2 + 5\gamma_{as}^2}{45\bar{\alpha}^2 + 4\gamma_s^2}, \tag{8}$$

where $\bar{\alpha}^2$ is the isotropic part of the Raman tensor:

$$\bar{\alpha}^2 = \frac{1}{9}\left(\sum_\rho \alpha_{\rho\rho}\right)^2,$$

γ_s^2 is the symmetric anisotropy

$$\gamma_s^2 = \frac{1}{2}\sum_{\rho,\sigma}(\alpha_{\rho\rho} - \alpha_{\sigma\sigma})^2 + \frac{3}{4}\sum_{\rho,\sigma}(\alpha_{\rho\sigma} + \alpha_{\sigma\rho})^2,$$

and γ_{as}^2 is the antisymmetric anisotropy

$$\gamma_{as}^2 = \frac{3}{4} \sum_{\rho, \sigma} (\alpha_{\rho\sigma} - \alpha_{\sigma\rho})^2.$$

In nonresonant scattering the Raman tensor is symmetric [the two terms of Equation (2) have reversed polarizations] i.e., $\alpha_{\rho\sigma} = \alpha_{\sigma\rho}$. Consequently, $\gamma_{as}^2 = 0$, and ρ cannot exceed $\frac{3}{4}$. Non-totally-symmetric modes have zero diagonal tensor elements, and give depolarized bands, $\rho = \frac{3}{4}$. Totally symmetric modes have $\bar{\alpha}^2 \neq 0$ and give polarized bands, $\rho < \frac{3}{4}$. In RR scattering, however, the tensor need not be symmetric. If $\gamma_{as}^2 \neq 0$, a band may be anomalously polarized, $\rho > \frac{3}{4}$. All three tensor invariants, $\bar{\alpha}^2$, γ_s^2, and γ_{as}^2, can be determined if circular as well as linear polarization measurements are made [41, 42].

A_{2g} modes have antisymmetric Raman tensors [41]; the only elements are $\alpha_{xy} = -\alpha_{yx}$. Thus, $\alpha^2 = \gamma_s^2 = 0$ and $\gamma_{as}^2 = 3\alpha_{xy}^2$. The depolarization ratio is expected to be infinite (inverse polarization [11]), since $I_{\parallel} = 0$. These modes have rotational symmetry ($\alpha_{xy} - \alpha_{yx}$ transforms as R_z), and they couple the x-component of the B electronic transition with the y-component of the Q electronic transition, and vice versa, thereby rotating the polarization plane of the scattered light by 90°. The high-frequency A_{2g} modes consist of radial C—H in-phase bending, or of alternate bond stretching and compression around the ring [43]. The bond-alternant modes are expected to be effective in vibronic mixing [39]. It is of interest that vibronic mixing via bond-alternant modes in aromatic annulenes is expected only for dianions, with $4n$ rather than $4n + 2$ atoms. The reason for this is evident from Simpson's model [25]. The two lowest π-π^* transitions have $\Delta l = 1$ and $\Delta l = 2n + 1$, and they can be coupled via vibrational modes having $2n$ units of angular momentum, i.e. $2n$ nodes. Bond-alternant modes change phase between every pair of atoms, and the number of nodes is therefore half the number of atoms. Consequently, $2n$ nodes require $4n$ atoms in the ring, rather than $4n + 2$. It follows that neutral aromatics, such as benzene, cannot be expected to show Raman activity for their bond-alternant modes, even at resonance.

For the A_{2g}-modes of many metalloporphyrins, $I_{\parallel}$ really is zero, within experimental error, but in other cases, e.g., cytochrome c [11], appreciable $I_{\parallel}$-intensities are seen. Circular-polarization measurements [42] showed significant values for $\bar{\alpha}^2$ and γ_s^2, as well as γ_{as}^2, consistent with lowering of the 4-fold porphyrin symmetry. Mixed polarization can be produced by protein-induced distortion, or by asymmetric peripheral substituents on the porphyrin ring. Warshel [44] has shown that the parallel scattering component induced into the 1580-cm^{-1} A_{2g} mode of Cu^{II} porphine by attachment of methyl groups at the 1, 3, 5, and 7 ring positions [45] is attributable simply to their kinematic-mass effect.

All of the porphyrin modes come into resonance at the Q_0 maximum (0-0) [46] and then again at the Q_1 position expected for the vibration being monitored (0-1) [11] (Figure 5). For B-term scattering equal enhancement is

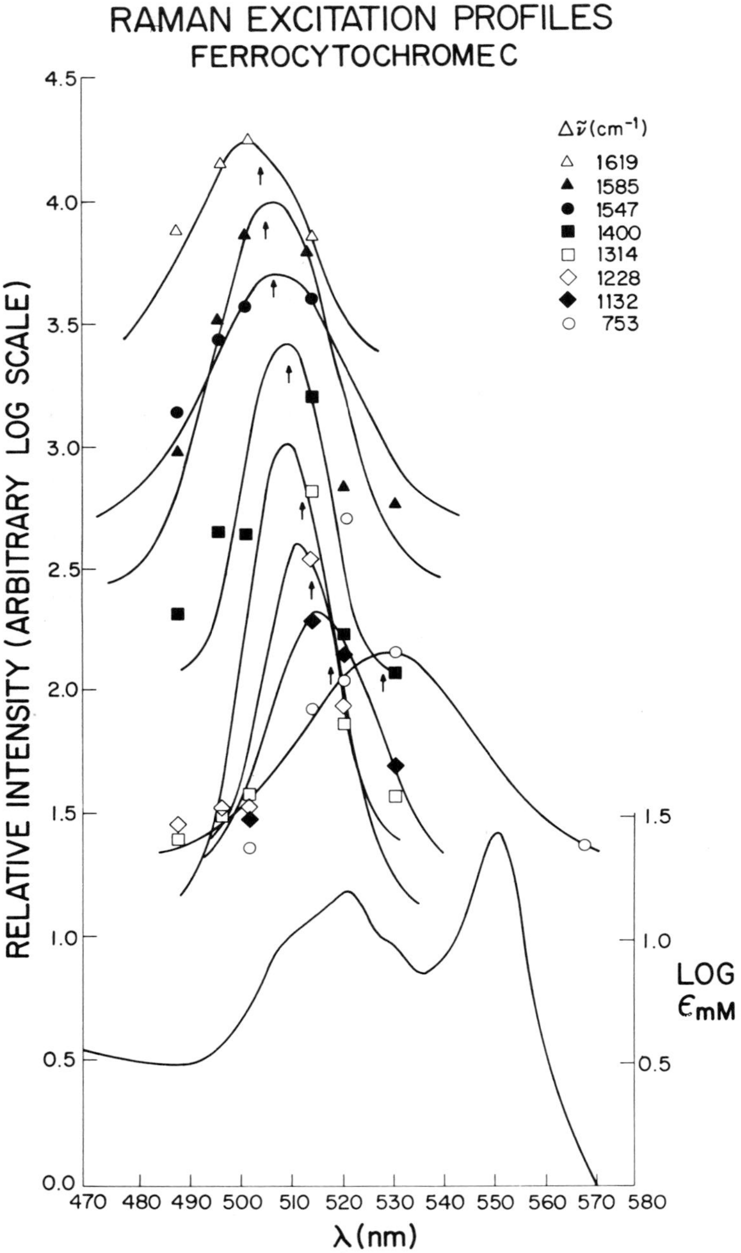

Figure 5 Excitation profiles for various RR bands of reduced cytochrome c in the region of the Q_1-band (absorption spectrum shown at the bottom) showing the expected shift of the resonance to higher energy with increasing vibrational frequency; the arrows mark the expected 0-1 positions, assuming no change in vibrational frequency in the excited state. The lines through the points are only for illustrative purposes and are not calculated. (Figure from [11].) The profile of the 753-cm^{-1} band has recently been shown [234] to be double-peaked, with 0-0 and 0-1 maxima.

expected at resonance with the 0-0 and 0-1 transitions, since the numerators of the first two terms in Equation (7), $\langle 1|Q|0\rangle\langle 0|0\rangle$ and $\langle 1|1\rangle\langle 1|Q|0\rangle$, are equal in magnitude. They have the same sign for symmetric (B_{1g} and B_{2g}) and opposite sign for antisymmetric (A_{2g}) modes [5]. Since the frequency denominator of Equation (7) changes sign as the incident frequency passes through a resonance, the two terms interfere destructively for B_{1g} and B_{2g} modes but constructively for A_{2g} modes, at frequencies between the 0-0 and 0-1 transitions. For lower symmetries, modes of mixed polarization show a dispersion of the depolarization ratio, which maximizes at $\nu_0 = (\nu_{00} + \nu_{10})/2$ [21], as has been observed for porphyrins (Verma, Mendelsohn, and Bernstein [27]). (Depolarization dispersion can also reflect splitting of the x and y components of the Q-band [47]). Off resonance, the terms cancel for antisymmetric vibrations at wavelengths outside the absorption band [46]. This cancellation is the reason that anomalous polarization of vibrational bands is only seen under rigorous resonance conditions.

In actual Q_0 and Q_1 excitation profiles, the 0-0 and 0-1 maxima are not always of equal height. Higher 0-1 maxima have been attributed to non-adiabatic effects [48], whereby extra enhancement at higher vibrational levels is predicted via nuclear-electron coupling with higher electronic states [48, 49]. Higher 0-0 maxima have been attributed to Jahn-Teller activity in the degenerate Q-state [49]. If a mode is active in both interstate and intrastate (Jahn-Teller) coupling, then there can be interferences between the two mechanisms that affect the excitation profiles [37, 50].

iii. Out-of-Plane Enhancement

a. Porphyrin Out-of-Plane Deformations. The out-of-plane modes of a planar metalloporphyrin involve bending of the in-plane bonds, and are all expected at low frequencies, < 1000 cm^{-1}. In D_{4h} symmetry, they span the representations $A_{1u}, A_{2u}, B_{1u}, B_{2u}$, and E_g.

Of these, only the E_g modes are Raman-active. There is no mechanism for E_g-enhancement via the in-plane electronic transitions alone. They can, however, be enhanced by vibronic mixing of in-plane (E_u) and out-of-plane (A_{2u}) electronic transitions: $E_u \times A_{2u} = E_g$. An example of this enhancement mechanism has recently been found [51] in the RR spectrum of (ImH)$_2$Fe^{II}PP (ImH = imidazole, PP = protoporphyrin IX). When excited at 4579 Å, this complex shows a band at 841 cm^{-1} (Figure 6) which disappears upon deuteration of the methine carbon atoms of the porphyrin ring, and which is assignable to out-of-plane bending of the methine C—H bonds [52]. From oriented-crystal spectroscopy, low-spin Fe^{II} hemeproteins are known to have weak z-polarized (A_{2u}) absorptions near 4579 Å, which are assigned [24] (Chapter 3) to porphyrin (a_{2u}) → Fe(d_{z^2}, a_{1g}) charge-transfer (CT) transitions (see Figure 10, and the discussion in Section B.iii.b.) The A_{2u} orbital concentrates electron density on the methine carbon atoms [53], C_m, and it is

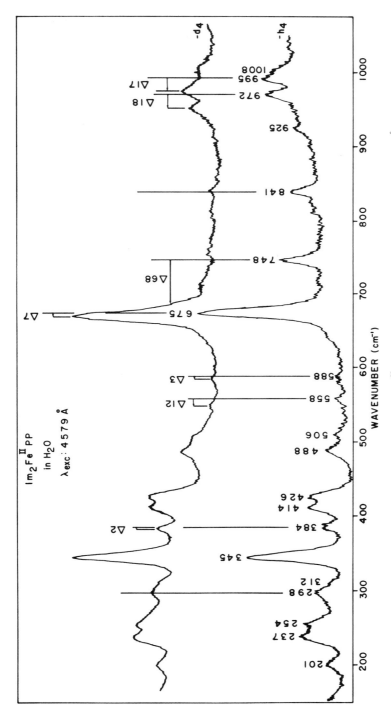

Figure 6 Low-frequency RR spectra of aqueous $(ImH)_2Fe^{II}PP$ and its $meso$-d_4 isotopic form with 4579-Å excitation. Note the disappearance of the 841-cm^{-1} band, assigned to γC_m—H out-of-plane bending. (From [51].)

suggested [51] that the $E_g \gamma C_m$—H mode is effective in mixing the out-of-plane CT transition with the nearby in-plane B-transition.

The resonance is quite local, since the 841-cm^{-1} band is not seen when excitation is shifted to nearby wavelengths (4880 or 4131 Å). Moreover, it only appears for aqueous solutions of the complex, in which the B absorption is split into two components (Figure 7) via molecular aggregation (which does not, however, alter the Raman frequencies significantly [51]). One of these components, at 435 nm, is in close proximity with the laser wavelength and the position of the CT transition (which is too weak to see in the unpolarized absorption spectrum [24]). When the complex is dissolved in methanol, the B-absorption collapses to a single peak at 413 nm and the 841-cm^{-1} Raman band disappears.

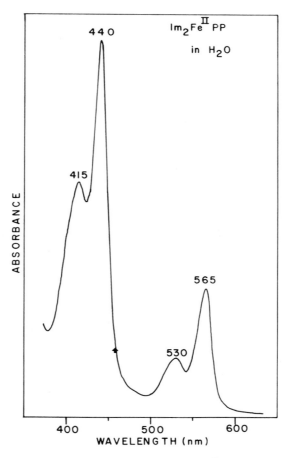

Figure 7 Absorption spectrum of aqueous $(ImH)_2 Fe^{II}PP$, showing the split B-band, due to aggregation, and the excitation wavelength (arrow) for the spectra in Figure 6.

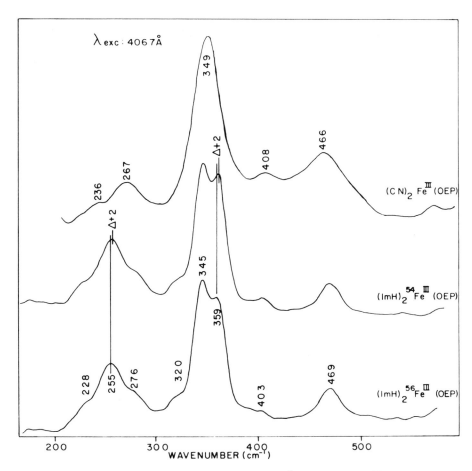

Figure 8 Low-frequency RR spectra ($\lambda_0 = 4131$ Å) of [(ImH)$_2$Fe^{III}OEP]Cl, and of its ^{54}Fe isotopic form, as well as of K[(CN)$_2$Fe^{III}OEP], all in dimethylformamide (~ 1 mM). Note the isotope shifts of the bands at 255 and 359 cm^{-1}, and their disappearance, as well as the disappearance of the 320-cm^{-1} band, in the cyanide complex. (From [51].)

Figure 8 [51] shows evidence for a different enhancement mechanism associated with out-of-plane modes at 255 and 359 cm^{-1} in [(ImH$_2$Fe^{III}OEP]$^+$ (OEP = octaethylporphyrin), observed when its RR spectrum is obtained with B-excitation. Both of these bands show a 2-cm^{-1} upshift upon ^{54}Fe substitution, and they disappear when the ImH ligands are replaced by CN$^-$. Since the Fe atom is at the symmetry center of the molecule, any isotope shift must be associated with i.r. modes, $E_u(x, y)$ or $A_{2u}(z)$. Activation of E_u modes is implausible, since the peripheral substituents are identical in OEP, and maintain the center of symmetry in the porphyrin plane; none of the E_u ring modes are seen for $[(ImH)_2Fe^{III}OEP]^+$, in contrast to protoporphyrin complexes, in which the in-plane symmetry is lost (see Section C.ii.a). The imidizole ligands are likewise unable to move the Fe atom off center within the plane, unless the Fe—ImH bonds are tilted; the actual tilt observed in the crystal structure [54] of $[(ImH)_2Fe^{III}TPP]^+$ is only $\simeq 3.5°$ from the porphyrin normal.

The same crystal structure, however, shows a pronounced inequivalence of the two imidazole ligands [54]. The Fe—ImH bond distances, 1.957 and 1.991 Å, differ significantly, and the longer bond is associated with an orientation of the ImH ring that brings its H atoms into close nonbonded contacts ($\simeq 2.60$ Å) with two of the pyrrole N atoms. RR activation of A_{2u} out-of-plane modes could well arise from this kind of ImH inequivalence, especially as the two bands in question are plausibly assigned to coupled pyrrole-tilting and peripheral substituent deformation modes (see Section C.i.b). The ImH inequivalence in the $[(ImH)_2Fe^{III}TPP]^+$ crystal structure can be related to differential H-bonding of the ImH N1 protons to the Cl^- counterion and to a cocrystallized methanol molecule [54]. In solution, there is no doubt a great deal of conformational flexibility. The coupling between ImH orientation and the Fe—ImH distance, via the nonbonded contacts, as well as the H-bonding opportunities (the solution contains an excess of ImH), may nevertheless lead to asymmetrical structures on the average. The disappearance of these 255- and 359-cm^{-1} bands upon replacing ImH with the less sterically demanding CN^- ligands strongly supports this hypothesis [51].

A third band in the $[(ImH)_2Fe^{III}OEP]^+$ RR spectrum, at 320 cm^{-1}, also disappears when ImH is replaced by CN^-. This band shifts down by 10 cm^{-1} when the methine carbon atoms are deuterated, and is assigned to out-of-plane deformation of the methine bridges (see Section C.i.b). Its activation in parallel with the 255- and 359-cm^{-1} A_{2u} modes likewise suggests an A_{2u} assignment. It is no doubt the same mode as was earlier observed at $\simeq 330$ cm^{-1} in RR spectra of $Mn^{III}EP$ (EP = etioporphyrin) halides by Asher and Sauer [55], and assigned as out-of-plane because its intensity increased with increasing size of the halide ($F^- < Cl^- < Br^- < I^-$), and, presumably, with the out-of-plane displacement of the Mn^{III} ion. Also, its frequency shifted up slightly (~ 1 cm^{-1}) on $^{37/35}Cl^-$ substitution, indicating coupling with the Mn—Cl stretching mode (286 → 282 cm^{-1}) [54]. In FePP complexes as well, the $\simeq 320$-cm^{-1} band intensifies for five-coordinated out-of-plane structures [51].

Out-of-plane assignments have been suggested for $Mn^{III}Mb$ RR bands at 170 and 280 cm^{-1}, which are activated upon azide addition [56]. In $Fe^{III}Mb$, RR bands in the $\simeq 450$-cm^{-1} region which shift down upon vinyl deuteration are assignable to pyrrole-folding modes [51] (see Section C.i.b). In hemeproteins, direct coupling of out-of-plane modes to the π-π^* transitions can result from loss of the porphyrin mirror plane via asymmetric ligation, protein-induced distortion of the porphyrin ring [57], or asymmetric electrostatic fields in the heme-binding region [58].

b. Axial-Ligand Modes

π-π* *Resonance.* Metalloporphyrins can bind one or two ligands at the axial coordination sites of the central metal ion. Since the chemistry of the heme group is determined by the nature of the axial ligands, it has been a matter of great interest to identify the metal-ligand stretching modes in the RR spectra. Although these modes are generally weak, it has proved possible to assign a number of them, using isotopic frequency shifts (see Section C.iii.a). It is important to clarify the enhancement mechanisms available to these modes, in order to find optimal conditions for detecting them. This task is still at a preliminary stage.

It has been conventional wisdom that the in-plane porphyrin electronic transitions do not enhance M—L stretching modes, since the latter are out-of-plane vibrations. Unlike the porphyrin out-of-plane deformations, however, the M—L stretches are not restricted by symmetry from coupling to the in-plane electronic transitions. If a D_{4h} metalloporphyrin has two identical axial ligands, the M—L_2 symmetric stretch has A_{1g} symmetry (the asymmetric stretch, A_{2u}, is infrared but not Raman-active, unless the mirror plane is destroyed). If the ligands are not the same, or if there is only one axial ligand, the point group is lowered to C_{4v}, and the M—L stretch remains A_1. Thus a M—L stretching mode can be enhanced by an in-plane transition, provided that the excitation alters the M—L bond length.

That this can be an acceptable mechanism is demonstrated by the excitation profile for the Fe—O_2 stretching mode of HbO_2 (oxyhemoglobin) shown in Figure 9 [59]. Similar observations have been made by Tsubaki et al. [68]. This mode is enhanced strongly in the B-band, and more weakly in the region of the Q-bands. The coupling of νFe—O_2 to the in-plane π-π* transitions can be understood on the basis of the electronic structure of oxy heme. O_2 is a π-acceptor ligand, and competes effectively with the porphyrin ring for backbonding of the iron d_π electrons [43]. When poor π-acceptors, such as ImH, are bound to low-spin Fe^{II} hemes, a number of porphyrin ring frequencies are strongly perturbed (see Section D.ii) as a result of d_π back donation into the low-lying porphyrin π* orbitals [43, 60]. These shifts are reversed upon removal of an electron, leaving low-spin Fe^{III}, or upon replacement of ImH with a good π-acceptor ligand [61], such as O_2 or NO [62]. It follows that the direction of backbonding, and therefore the energy of the porphyrin π*-orbitals, should be sensitive to the Fe—O_2 bond length, and that Fe—O_2 should modulate the π-π* transitions.

The Fe—O_2 stretch, identified via its $^{13}O_2$ shift, was the first axial mode to be discovered [63]. Its enhancement was expected [16 (1981)] to be due to z-polarized transitions, which have been identified in the MbO_2 crystal spectrum [24, 64]. The z-polarized band at $\simeq$ 475 nm, close to the excitation wavelength initially used for the study of νFe—O_2 [63], has been assigned to either $O_2 \rightarrow$ Fe CT [65] or porphyrin $\pi(a_{2u}) \rightarrow$ Fe $d_{z^2}(a_{1g})$ CT [66]. Either

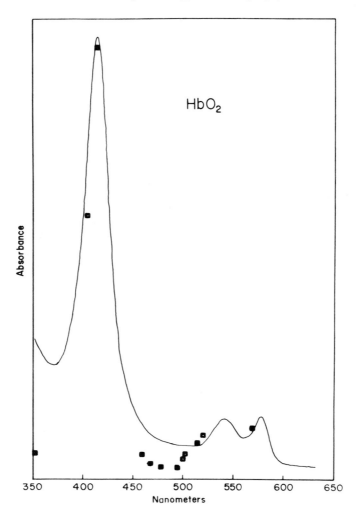

Figure 9 Excitation profile (filled squares) of the Fe—O_2 stretching mode in HbO_2, superimposed on the absorption spectrum (from [59]).

transition should enhance νFe—O_2, since either the bonding O_2 or the antibonding Fe(d_{z^2}) orbitals are involved. The excitation profile, however, is flat in the region of this absorption, and the conclusion seems inescapable that CT enhancement must be less than or comparable to the preresonance enhancement from the *B*-transition, presumably because the CT transition moment is small.

A similar π-π* enhancement mechanism can be expected for the stretching mode of other π-acceptor ligands. Indeed, the Fe–NO stretch of HbNO, discovered via its $^{15/14}$NO shift by Chottard and Mansuy [67], also shows intensification in the B and Q_1 bands, and has a flat profile in between [67]. Recently the Fe—CO stretch of CO-hemoglobin has been found to be quite strongly enhanced in the B-band [68]. Moreover, the Fe—C—O band, and the C—O stretch, although weaker, are also enhanced (Tsubaki et al. [68]).

The Fe—(ImH)$_2$ symmetric stretch has been identified at 200 cm^{-1} for the bis-imidazole protoporphyrin IX complexes of both Fe^{II} and Fe^{III}, from the shifts induced upon imidazole perdeuteration [69a], and for Fe^{II}, by the imidazole ^{15}N shift [69b]. For the Fe^{II} complex this mode, although weak, shows observable enhancement upon direct B-band excitation. ImH is not an effective π-acceptor, but stretching of the Fe—ImH bond should nevertheless lower the $d\pi$ orbital energies, via polarization, and therefore modulate the strong Fe^{II} back donation to the porphyrin e_g-orbitals [43, 60, 61]. In the case of Fe^{III}, however, the mode is not seen with B-excitation, but only with 4579 Å in aqueous solution, at the edge of the (aggregation-induced) split B-bond [69]. The enhancement mechanism is uncertain; it may involve vibronic coupling (see above). But the lack of direct π-π^* enhancement is consistent with the diminished π-backbonding in Fe^{III} hemes.

In deoxyHb or Mb, or in the analog complex (2-MeImH)Fe^{II}PP, the Fe—ImH stretch, identified near 200 cm^{-1} via 2-MeImH perdeuteration [70] and $^{54/56}$Fe substitution [71], stands out quite strongly upon excitation near the B-band. In this five-coordinated heme, the Fe atom is out of the porphyrin plane by $\simeq 0.5$ Å [72], and when the Fe—ImH bond is stretched, the Fe atom moves toward the plane, thereby altering the interaction of the Fe orbitals with the porphyrin π and π^* orbitals. This provides another mechanism for π-π^* coupling. A similar argument applies to other five-coordinated hemes, e.g. ν Fe—F in (F$^-$)Fe^{III}MP [73] (MP = mesoporphyrin IX), ν Fe—X (X = F, Cl, Br) in (X$^-$)Fe^{III}OEP [74], ν Fe—O—Fe in O(Fe^{III}TPP)$_2$ [35].

Kitagawa et al. [74] have suggested that vibrational coupling with low-frequency in-plane porphyrin modes might account for axial-mode enhancements. Vibrational coupling is expected to be weak, however, since the internal coordinates are orthogonal, or nearly so. If the metal atom is in the plane, then there is no kinematic coupling between the axial stretches and the in-plane displacements; vibrational coupling depends on axial–in-plane interaction constants, which are expected to be small. It seems probable that direct coupling of the axial stretches to the in-plane electronic transitions is a more important factor in the enhancement.

Change-Transfer Resonance

Porphyrin → Fe CT Because transition-metal ions have partially filled d-orbitals, a variety of charge-transfer transitions are possible for their complexes with porphyrins. Figure 10 is a qualitative orbital energy diagram, which includes a set of d-orbitals and the four porphyrin frontier orbitals that dominate the visible and near-u.v. spectrum. The relative ordering of these orbitals has been examined [26] for a wide range of complexes, using extended Hückel calculations and the available spectroscopic data. The center of gravity

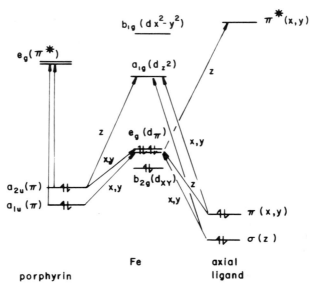

Figure 10 Schematic diagram of the frontier orbitals of a low-spin $Fe^{(II, III)}$ porphyrin, with axial ligands. The arrows represent allowed transitions, with the indicated polarizations.

of the d-orbitals drops with increasing effective nuclear charge on the metal, from left to right across the period table, and from lower to higher oxidation states.

The splitting of the d-orbitals depends on the ligand field of the porphyrin and the axial ligands. The strong in-plane porphyrin field drives the $d_{x^2-y^2}$ orbital up, usually above the porphyrin e_g^* orbitals. The position of the d_{z^2} orbital is highly variable, depending on the axial ligands. Strong-field ligands, such as NO or CN^-, drive it up to or beyond $d_{x^2-y^2}$, while for weak field or absent axial ligands it falls toward the d_π (d_{xz} and d_{yz}) and d_{xy} orbitals. The position of the d_π orbitals, relative to the nonbonding d_{xy}, is influenced by π-bonding. They are stabilized by π-acceptor ligands such as NO and O_2 and by the porphyrin ring, and destabilized by π-donor ligands such as OH^- and halides.

Charge transfer can take place from the filled porphyrin orbitals, a_{2u} and a_{1u}, to d-orbital vacancies, or from occupied d-orbitals to the vacant porphyrin orbitals, e_g^*. The latter process is, however, parity-forbidden. This restriction can be removed by mixing in metal p-orbital character, for complexes lacking inversion symmetry, but it is unlikely that $d \to e_g^*$ transitions can contribute significantly to RR enhancement. The $a_{2u} \to d_{z^2}$ transition is allowed ($a_{1u} \to d_{z^2}$ and $a_{1u}, a_{2u} \to d_{x^2-y^2}$ are not), but the orbital overlap is poor. The transition is weak, but it can be identified in polarized-crystal spectra via its z-polarization, and several $a_{2u} \to d_{z^2}$ assignments have been suggested [24]. Intensification (and RR enhancement) via vibronic mixing with the intense π-π^* transitions requires out-of-plane, E_g mixing modes, as discussed above.

The $a_{2u}, a_{1u} \to d_{xy}$ transitions are forbidden, but $a_{2u}, a_{1u} \to d_\pi$ transitions are allowed, and are polarized in the plane. These have been assigned near 600 nm for high-spin Fe^{III} and around 1200 to 1500 nm for low-spin Fe^{III} [24, 26]. The large reduction in transition energy between high- and low-spin Fe^{III} is attributable to interelectronic-repulsion effects [26]. The $a_{2u}, a_{1u} \to d_\pi$ transitions have not been identified in the spectra of high-spin Fe^{II} porphyrins [24]. For low-spin Fe^{II}, they are necessarily absent, since the d_π orbitals are completely filled.

When the d_π orbitals approach the e_g^* orbitals in energy, strong mixing is expected between the $a_{2u}, a_{1u} \to d_\pi$ and the $a_{2u}, a_{1u} \to e_g$ (B and Q) transitions, which have the same symmetry, E_u. Strong extra absorptions appear in the resulting "hyperporphyrin" spectra, which Gouterman has analyzed extensively [26]. Mn^{III} porphyrins are notable exemplars of such spectra, showing a split "Soret" absorption [75] presumably resulting from a near coincidence of d_π and e_g^* energy levels. The lower-energy component, labeled band V [75], falls at $\simeq 460$ nm and is convenient for RR studies with Ar^+ laser excitation. Several groups [55, 56, 76, 77] have reported Mn^{III} porphyrin RR studies, and arguments have been advanced for predominant π-π^* [77] or CT [55] character of the band-V transition. In view of the high intensity of both Soret components, it seems evident that the π-π^* and CT transitions are heavily admixed. Both porphyrin skeletal and metal-ligand modes are observed, but it is clear from the spectra of Asher and Sauer [55] that the low-frequency modes, including Mn—X (X = halide) stretches and in-plane modes that probably involve Mn—pyrrole stretching, are much stronger, relative to the high-frequency modes, than is usually the case for B-band excitation of metalloporphyrins. This intensification is plausibly linked to the CT involvement [55] since stretching of both Mn—X and Mn—pyrrole bonds should lower the energy of the d_π orbitals. Low-frequency modes are also prominent in the RR spectra of Cr^{III} porphyrins [77], which likewise give rise to hyperporphyrin spectra [26], presumably reflecting strong CT mixing.

As noted above, absorption bands near 600 nm have been attributed to $a_{2u}, a_{1u} \to d_\pi$ transitions of high-spin Fe^{III} hemes. They are comparable in intensity to the Q-bands, with which they are no doubt strongly mixed. Asher et al. [78] reported specific enhancement of Fe–axial-ligand modes between 400 and 500 cm^{-1} for methemoglobin (Hb^{III}) complexes of fluoride, hydroxide, and azide. The Fe—OH assignment, at 497 cm^{-1} in $Hb^{III}OH$ and 490 cm^{-1} in $Mb^{III}OH$ [79], was confirmed by its $^{18}OH^-$ shift [78, 80], and Desbois et al. [80] showed this band, as well as one of the pair (462 cm^{-1}) of bands

(462 and 422 cm^{-1}) in Mb^{III}F assigned by Asher and Schuster [79] to Fe—F stretching, to shift upon ^{54}Fe substitution. Later Tsubaki et al. [81] confirmed by ^{15}N$_3$ substitution that the 411-cm^{-1} band in Mb^{III}N$_3$ (413 cm^{-1} in Hb^{III}N [78]) was indeed due to Fe—N$_3$ stretching, but they also showed, via its temperature dependence, that this band was associated with the low-spin rather than the high-spin component of Mb^{III}N$_3$. Therefore its enhancement at 6471 Å could not be due to the $a_{2u}, a_{1u} \rightarrow d_\pi$ CT transitions, which are at much lower energy for low-spin Fe^{III} hemes. It was attributed instead to a N$_3^- \rightarrow$ Fe^{III} (low-spin) CT transition (see next section).

A similar assignment has to be considered for the Fe—OH band of Hb^{III}OH and Mb^{III}OH, which also contain spin mixtures, but Asher and Schuster [79] showed that the Fe—OH excitation profile of Mb^{III}OH tracked those of the high-spin porphyrin skeletal modes, at 1545 and 1608 cm^{-1}, while the low-spin mode at 1644 cm^{-1} peaked at higher energy, $\simeq 580$ nm, where the low-spin Q-band is expected. Thus enhancement via the high spin $a_{2u}, a_{1u} \rightarrow d_\pi$ transitions seems likely. Hb^{III}F and Mb^{III}F contain purely high-spin hemes, and the enhancement via the strong $\simeq 600$-nm band [79] must be attributed to the $a_{2u}, a_{1u} \rightarrow d_\pi$ transition (F$^- \rightarrow$ Fe^{III} CT being most unlikely at this low an energy); the Q_0 band has been assigned [28, 82] at $\simeq 528$ nm. Tsubaki et al. [81] argue, however, that Fe—ligand modes are unlikely to be strongly involved in the $a_{2u}, a_{1u} \rightarrow d_\pi$ excited states and point to the lack of an identifiable Fe—ligand stretch in the RR spectrum of high-spin acid ferricytochrome c, when excited in its 620-nm absorption band [83].

Ligand $\rightleftharpoons$ M CT

Charge-transfer transitions are also possible to or from the axial ligands of metalloporphyrins. The situation is shown diagramatically on the right side of Figure 10. Transitions are possible from filled π or σ orbitals on the ligands to vacancies in any of the d-orbitals (L $\rightarrow$ M CT), or from (partially) filled d-orbitals to π^*-orbitals on the ligands (M $\rightarrow$ L CT). All of these transitions are allowed (strictly speaking, only one of the linear combinations of each pair of σ, π, or π^* orbitals on a symmetric ML$_2$ unit can take part in an allowed transition), but their intensities will vary with the extent of orbital overlap. Highest intensities are expected for L(σ) $\rightarrow$ M(d_{z^2}), L(π) $\rightarrow$ M(d_π), and M(d_π) $\rightarrow$ L(π^*) transitions. The energies of these transitions are not easy to anticipate, since the d-orbital energies are influenced by their interactions with the porphyrin and with the axial ligands in a concerted manner. From the properties of simple complexes, it is expected that for Fe^{III}, L $\rightarrow$ M CT transitions will occur in the visible region for oxidizable ligands such as Br$^-$, RS$^-$ and N$_3^-$, while visible region M $\rightarrow$ L CT transitions can occur from low-spin Fe^{II} to ligands with low-lying π^*-orbitals. For many ML combinations, the transitions shown in Figure 10 lie in the near or far u.v.

Because metalloporphyrin absorption spectra are dominated by the intense, broad π-π^* transitions, it is difficult to locate L $\rightarrow$ M or M $\rightarrow$ L CT absorptions, and the few cases where they are believed to provide RR enhancements were found serendipitously. Thus early studies of bis-pyridine(py) Fe^{II} porphyrins ("hemochromes") showed strong enhancement of internal modes of the bound pyridine ligands [61]. These modes, as well as the py—Fe—py symmetric stretch, were identified via their frequency shifts upon perdeuteration of py [84], and their excitation profiles were found to coincide with a bump in the absorption spectrum, which had been suggested to arise from $Fe^{II} \rightarrow$ py CT [85]. Consistent with this assignment, an analysis of the relative intensities indicated a pattern of bond-length changes in the excited state similar to that expected for occupation of the lowest pyridine π^*-orbital [84].

Yu and Tsubaki [56] observed RR bands at 2039 and 650 cm^{-1} for the azide complex of Mn^{III}-substituted Mb, whose strong $^{15}N_3^-$ shifts implicated them as internal azide modes. They were both depolarized, implying that the modes are not totally symmetric with respect to the local (linear) N_3^- geometry (the complex as a whole has only mirror symmetry, since the MnN_3 unit is undoubtedly bent, as in other azides). They were assigned as the antisymmetric N_3^- stretch and the in-plane N—N—N bend. As with other Mn^{III} porphyrins, $Mn^{III}Mb(N_3^-)$ shows strong absorptions at $\simeq$ 380 and 470 nm, due to mixed $a_{2u}, a_{1u} \rightarrow d_\pi, e_g^*$ transitions (see above), but the internal azide modes were found to be enhanced *between* these bands, implying enhancement via an additional CT transition involving the azide; the enhancement mechanism was attributed to vibronic mixing between this extra CT transition and the intense in-plane transitions. Yu and Tsubaki [56] suggested that the extra CT transition was $N_3^-(\pi) \rightarrow$ porphyrin(π^*) rather than $N_3^-(\pi) \rightarrow Mn(d_{z^2})$ or $N_3^-(n) \rightarrow Mn(d_{z^2})$ in character, since no band attributable to Mn—N_3^- stretching was observed. In view of the probable proximity of porphyrin e_g^* and Mn^{III} d_π-orbitals, $N_3^-(\pi) \rightarrow Mn(d_\pi)$ character is also likely. Lack of Mn—N_3^- enhancement via such a transition would not be surprising on the precedent of the transferrin RR spectrum [86], which shows no strong Fe^{III}—ligand modes upon excitation into a phenolate $\rightarrow Fe^{III}(d_\pi)$ CT band, although phenolate-ring modes are strongly enhanced; the excitation was suggested [86] not to alter the Fe—O bond length appreciably because of the nonbonding character of the initial (phenolate π) and final (Fe^{III} d_π) orbitals.

The Fe^{III}—N_3^- mode is seen in the azide complex of Fe^{III} myoglobin, as mentioned above, and was shown by Tsubaki et al. [81] to be due to the low-spin form. It is enhanced in the vicinity of 600 nm [81], and since no porphyrin $\rightarrow$ low-spin Fe^{III} CT transition is expected at this wavelength, a $N_3^- \rightarrow Fe^{III}$ CT transition may be responsible. But another $N_3^- \rightarrow Fe^{III}$ CT transition seems definitely to be located near 4067 Å, the position of the B-band, since at this wavelength the RR spectrum revealed the antisymmetric N_3^- stretch at 2024 cm^{-1} (low-spin form) and a depolarized, $^{15}N_3^-$-sensitive band, assigned to the out-of-plane N—N—N bend [81]. (It had previously been thought to be the Fe—N_3 stretch [80 (1979)]. Again vibronic mixing of the CT and B-bands was invoked to explain the depolarized enhancement. Tsubaki et al. assigned the $\simeq$ 400-nm CT transition to $N_3^-(\pi) \rightarrow Fe^{III}(d_{z^2})$

and the $\simeq$ 600-nm CT transition to $N_3^-(n) \rightarrow Fe^{III}(d_{z^2})$, on the grounds that enhancements of internal azide modes are expected for $N_3^-(\pi)$ but not $N_3^-(n)$ excitations, and none are seen with $\simeq$ 600-nm excitation. This assignment places an azide n-orbital $\simeq$ 8000 cm^{-1} above the π-orbital; this level ordering could presumably result from the sp^2 hybridization of the terminal N atom, associated with a bent Fe—N_3 unit, which would leave a relatively high-energy lone pair n-orbital.

Tsubaki and Yu [87] found that excitation of Co-substituted MbO$_2$ and HbO$_2$ at 4067 Å produces enhancement of both Co—O$_2$ and O—O stretching modes. Two ν_{O-O} frequencies are observed, suggesting different CoO$_2$ conformations. One of them interacts with a coincident porphyrin mode, producing a pair of bands, at 1103 and 1137 cm^{-1}; $^{18}O_2$ substitution decouples the interaction, shifting ν_{O-O} to 1069 cm^{-1}, and leaving the porphyrin band at 1123 cm^{-1} (probably ν_{44}, a C_b—vinyl stretch; see Section C.ii.a) with increased intensity. In view of the enhancement of both ν_{Co-O_2} and ν_{O-O}, resonance with a $O_2(\pi^*) \rightarrow Co(d_{z^2})$ CT transition near 400 nm was suggested. A similar transition was earlier assigned by Nakamoto et al. [88] to $\simeq$ 500-nm bands of the binuclear O$_2$ adducts [LCo(salen)]O$_2$ (L = pyridine, pyridine-N-oxide, and dimethylformamide; salen = N,N'-ethylenebis(salicylideniminate)] on the basis of RR enhancements of ν_{Co-O_2} and ν_{O-O} in these complexes. Because the higher O$_2$ π^*-orbital is empty in Fe^{II} porphyrin adducts, the analogous transition is absent for MbO$_2$ or HbO$_2$; Tsubaki and Yu did, however, find evidence for a similar ν_{O-O}-porphyrin mode interaction in the intensification of the MbO$_2$ 1125-cm^{-1} band upon $^{18}O_2$ substitution [87].

C. VIBRATIONAL ASSIGNMENTS

i. Porphyrin Skeletal Modes

a. In-plane Modes. The porphyrins found in nature have the basic porphine skeleton, with H atoms at the methine bridges, and carbon substituents at the pyrrole C_b atoms (see Figure 1 for the structure and atom labeling.) If, conceptually, these substituents are replaced by single masses S, and if the porphyrin is planar, with a metal atom at its center, then the resulting D_{4h} model has 71 in-plane modes, classified as

$$\Gamma_{\text{in-plane}} = 9A_{1g} + 8A_{2g} + 9B_{1g} + 9B_{2g} + 18E_u.$$

Eighteen infrared bands are expected, and 35 Raman bands, nine of them polarized (A_{1g}), 18 depolarized (B_{1g} and B_{2g}), and eight anomalously polarized (A_{2g}). The polarizations permit classification of the symmetry species of the Raman-active modes [11] and are a principal tool for assigning the complex vibrational spectrum.

There are four C_m—H stretching modes, $A_{1g} + B_{2g} + E_u$, with frequencies near 3000 cm^{-1}. No significant RR enhancement is expected, or observed,

Table 1. Internal Coordinate Contributions to the Normal Modes of a D_{4h} Metalloporphyrin with Point-Mass Pyrrole Substituents[a]

	A_{1g}	A_{2g}	B_{1g}	B_{2g}	E_u
$\nu C_m H$ [b]	1	0	0	1	1
$\nu C_b C_b$	1	0	1	0	1
$\nu C_a C_m$	1	1	1	1	2
$\nu C_a N$	1	1	1	1	2
$\nu C_a C_b$	1	1	1	1	2
$\nu C_b S$	1	1	1	1	2
νMN	1	0	1	0	1
$\delta C_a C_m H$	0	1	1	0	1
$\delta C_b C_b S$	1	1	1	1	2
$\delta C_a C_m C_a$	1	0	0	1	1
$\delta C_b C_b C_a$	1	1	1	1	2
$\delta C_b C_a C_m$ $\delta C_b C_a N$ $\delta C_m C_a N$	2	2	2	2	4
$\delta C_a NC_a$ $\delta C_a NM$	1	$\begin{Bmatrix}0\\1\end{Bmatrix}$	1	$\begin{Bmatrix}0\\1\end{Bmatrix}$	$\begin{Bmatrix}1\\2\end{Bmatrix}$
δNMN	1	0	0	1	1
	14	10	12	12	25

[a] See Figure 1 for atom labeling. In-plane vibrational representation $= 9A_{1g} + 8A_{2g} + 9B_{1g} + 9B_{2g} + 18E_u$; cyclic redundancies $= 5A_{1g} + 2A_{2g} + 3B_{1g} + 3B_{2g} + 7E_u$.
[b] $\nu XY = X-Y$ bond stretching; $\delta XYZ = X-Y-Z$ angle bending.

since stretching of the C—H bonds is not appreciably coupled to the π-π^* transitions. Modes involving C—C, C—N, and C—S stretching and C_m—H in-plane bending are expected at frequencies between 900 and 1700 cm^{-1}. This region contains the strongest RR bands. The contributions of these internal coordinate types to the various symmetry classes are given in Table 1. Below 900 cm^{-1}, the major contributors are angle deformations within the ring and of the C_b—S bonds, and stretching of the metal—N(pyrrole) bonds.

Using this model, Kitagawa et al. [89] assigned most of the in-plane modes of NiOEP, in which all eight pyrrole substituents are ethyl groups, using i.r. and variable excitation RR spectra, ^{15}N and C_m—D isotope shifts, and an analysis of RR combination modes. They carried out a normal-coordinate analysis [90], using a Urey-Bradley force field, which reproduced the assigned frequencies with reasonable accuracy. Because of the more complete assignments, this calculation superseded others which had earlier been reported [91–95]. A summary of the assignments and calculations is given in Table 2. Recently Warshel and Lappicirella [96] calculated metalloporphine mode frequencies, using a semiempirical force field (QCFF/PI), which are in fair agreement with the NiOEP results.

Table 2. Assigned and Calculated Normal-Mode Frequencies (cm^{-1}) for NiOEP [a]

Sym.	No.	Obs.	Δd_4 [b]	Δ^{15}N [c]	Calc.	Δd_4	Δ^{15}N	PED (%) [d]
A_{1g}	ν_1	—	—	—	3072	807	0	$\nu C_m H$ (100)
	ν_2	1602	0	0	1591	2	0	$\nu C_b C_b$ (60), $\nu C_b -$ Et (19)
	ν_3	1519	7	0	1517	4	1	$\nu C_a C_m$ (41), $\nu C_a C_b$ (35)
	ν_4	1383	1	6	1386	5	9	$\nu C_a N$ (53), $\delta C_a C_m$ (21)
	ν_5	1025	+1	3	1048	2	6	νC_b—Et (38), $\nu C_a C_b$ (23)
	ν_6	806	4	5	809	11	8	$\delta C_a C_m C_a$ (36), $\nu C_a N$ (27)
	ν_7	674	7	1	655	11	3	$\delta C_b C_a N$ (20), $\nu C_a C_b$ (19)
	ν_8	344	2	0	326	0	1	δC_b—Et (57), $\nu C_a C_m$ (11)
	ν_9	226	0	0	230	0	1	δC_b—Et (23), $\nu C_a C_m$ (16)
B_{1g}	ν_{10}	1655	10	0	1656	11	2	$\nu C_a C_m$ (49), $\nu C_a C_b$ (17)
	ν_{11}	1576	0	0	1587	0	0	$\nu C_b C_b$ (57), $\nu C_b -$ Et (16)
	ν_{12}	—	—	—	1351	5	10	$\nu C_a N$ (63), $\nu C_b C_b$ (13)
	ν_{13}	1220	270	0	1262	302	2	$\delta C_m H$ (67), $\nu C_a C_b$ (22)
	ν_{14}	—	(1187) [e]	—	1095	+50	4	$\nu C_a C_b$ (31), νC_b—Et (30)
	ν_{15}	—	—	—	754	3	11	νC_b—Et (25), $\nu C_a C_b$ (20)
	ν_{16}	751	67	2	741	72	2	$\delta C_a N C_a$ (14), νC_b—Et (14)
	ν_{17}	—	—	—	299	0	1	δC_b—Et (84)
	ν_{18}	—	—	—	187	0	0	$\delta C_a C_m$ (39), νMN (34)
A_{2g}	ν_{19}	1603	21	0	1600	16	1	$\nu' C_a C_m$ (67), $\nu' C_a C_b$ (18)
	ν_{20}	1397	0	1	1409	3	4	$\nu' C_a N$ (29), $\nu' C_b$—Et (24)
	ν_{21}	1308	418	3	1281	426	2	$\delta' C_m H$ (53), $\nu' C_a C_b$ (18)
	ν_{22}	1121	+81	13	1118	+87	8	$\nu' C_a N$ (37), $\nu' C_b$—Et (26)
	ν_{23}	—	(1029) [e]	—	1022	+33	1	$\nu' C_a C_b$ (26), $\nu' C_b$—Et (20)
	ν_{24}	739	6	7	723	7	11	$\delta' C_a C_m$ (43), $\delta' C_b$—Et (29)
	ν_{25}	—	—	—	528	6	1	$\delta' C_a C_b C_b$ (39), $\delta' C_b C_a N$ (17)
	ν_{26}	—	—	—	289	2	0	$\delta' C_b$—Et (41), $\delta' C_a C_m$ (31)
B_{2g}	ν_{27}	—	—	—	3072	807	0	$\nu' C_m H$ (100)
	ν_{28}	—	—	—	1469	6	3	$\nu' C_a C_m$ (52), $\nu' C_a C_b$ (21)
	ν_{29}	1409	1	1	1409	1	2	$\nu' C_a C_b$ (47), $\nu' C_b$—Et (26)
	ν_{30}	1159	0	9	1157	1	5	$\nu' C_b$—Et (49), $\nu' C_a N$ (28)
	ν_{31}	—	—	—	1016	20	17	$\delta' C_a C_m$ (25), $\delta' C_a C_m C_a$ (23)
	ν_{32}	785	0	0	789	0	0	$\delta' C_b$—Et (50), $\delta' C_a C_m$ (22)
	ν_{33}	—	—	—	536	3	2	$\delta' C_a C_b C_b$ (28), $\nu' C_b$—Et (15)
	ν_{34}	—	—	—	232	2	1	$\delta' C_a C_m C_a$ (25), δ'NMN (22)
	ν_{35}	—	—	—	182	1	1	$\delta' C_b$—Et (30), $\delta' C_a C_m$ (25)
E_u	ν_{36}	—	—	—	3074	803	0	$\nu C_m H$ (100)
	ν_{37}	1604	9	1	1633	12	1	$\nu C_a C_m$ (36), $\nu' C_a C_m$ (24)
	ν_{38}	1557	15	2	1592	3	0	$\nu C_b C_b$ (53), νC_b—Et (16)
	ν_{39}	1487	7	3	1498	10	3	$\nu' C_a C_m$ (39), $\nu' C_a N$(16)
	ν_{40}	1443	3	1	1454	3	2	$\nu' C_a C_b$ (40), $\nu' C_b$—Et (14)
	ν_{41}	1389	6	3	1362	4	10	$\nu C_a N$ (50), $\delta C_a C_m$ (11)
	ν_{42}	1268	93	2	1277	79	1	$\delta C_m H$ (57), $\nu' C_b$—Et (11)
	ν_{43}	1148	34	8	1139	41	6	$\nu' C_b$—Et (38), $\nu' C_a N$ (24)
	ν_{44}	1113	95	5	1079	65	4	νC_b—Et (29), $\nu C_a C_b$ (26)
	ν_{45}	993	50	7	1007	65	15	$\nu' C_a N$ (19), $\nu' C_a C_m$ (16)
	ν_{46}	924	81	3	895	65	3	$\nu' C_b$—Et (20), $\delta' C_a C_m$ (17)
	ν_{47}	726	4	7	741	7	10	νC_b—Et (27), $\nu C_a C_b$ (22)
	ν_{48}	605	8	3	620	19	1	$\delta' C_b$—Et (29), $\nu C_a C_m$ (13)
	ν_{49}	550	13	0	546	6	2	$\delta' C_a C_b C_b$ (29), $\delta C_a C_m C_a$ (14)

Table 2. *Continued*

Sym.	No.	Obs.	Δd_4 [b]	$\Delta^{15}N$ [c]	Calc.	Δd_4	$\Delta^{15}N$	PED (%) [d]
	ν_{50}	—	—	—	333	1	2	δC_b—Et (28), $\delta C_a C_m$ (15)
	ν_{51}	—	—	—	291	1	0	δC_b—Et (19), δNMN (18)
	ν_{52}	287	—	—	264	0	0	δC_b—Et (43), νMN (29)
	ν_{53}	—	—	—	178	0	0	$\delta'C_b$—Et (24), $\delta'C_a C_m$ (22)

[a] Reference [90].
[b] Δd_4: downshift (upshift indicated by +) upon substituting D for H at the four C_m atoms.
[c] $\Delta^{15}N$: downshift upon substituting ^{15}N for the four pyrrole N atoms.
[d] Percentage contributions (in parentheses) to the potential-energy distribution from the indicated coordinate: ν = stretch, δ = in-plane bend. ν' and δ' represent symmetry coordinates which are antisymmetric with respect to the C_2 axis of the pyrrole ring.
[e] Frequencies (cm^{-1}) of bands seen only for the $-d_4$ species.

The bond-stretching force constants [90] fell in the order $C_b C_b > C_a C_m > C_a N > C_b S > C_a C_b$, which is approximately the order of increasing bond lengths [97]. The high-frequency A_{1g} modes fall in the same order, with respect to the major contributors to the potential energy, although $C_a C_b$ stretching is mixed in as a minor contributor, and there are only four A_{1g} modes between 1000 and 1600 cm^{-1}. As mentioned earlier, the $C_a N$ breathing mode, at 1383 cm^{-1} in NiOEP, is the dominant RR feature upon excitation near the B-band. The mainly $C_b C_b$ modes are close in frequency among the A_{1g}, B_{1g}, and E_u blocks (1557–1602 cm^{-1}), but the mainly $C_a C_m$ modes vary widely, from 1519 cm^{-1} (A_{1g}) to 1650 cm^{-1} (B_{1g}), reflecting the more highly coupled nature of the methine bridge bonds. The A_{2g} and B_{2g} blocks have no $C_b C_b$ contributions, by symmetry, and their highest-frequency modes are $C_a C_m$ in character. The 1603-cm^{-1} A_{2g} mode, corresponding to the bond-alternant methine bridge stretch [43], is the dominant feature of the Q-band RR spectrum, and is anomalously polarized. The $\delta C_m H$ in-plane deformation coordinate contributes to the A_{2g}, B_{1g}, and E_u blocks, and is calculated to be the major contributor to modes at 1308, 1220, and 1268 cm^{-1}.

The role of metal—N (pyrrole) stretching in the low-frequency modes deserves some discussion. The four M—N bond stretches classify as $A_{1g} + B_{1g} + E_u$. If the pyrrole rings were not part of the macrocycle, then two Raman modes would be expected near 250 cm^{-1}, by analogy with tetragonal Cu(ImH)$_4$X$_2$ complexes [98], with the infrared mode at somewhat higher frequency. In fact, a mainly M—pyrrole E_u mode has been located at 204 cm^{-1} in the i.r. spectrum of ZnOEP [99], via its $^{64/68}$Zn shift, while Abe et al. [90] have calculated a major Ni—N contribution for a NiOEP E_u mode at 287 cm^{-1}. However, because the pyrrole rings *are* part of the macrocycle, the Raman M—N(pyrrole) stretches, A_{1g} and B_{1g}, are completely redundant with internal-angle deformations of the macrocycle. The addition of the central metal ion adds no Raman modes to those of the porphyrin dianion. Contributions from the M—N force constant can be expected for various low-frequency

A_{1g} and B_{1g} modes: The calculations of Abe at al. [90] show a major Ni—N contribution (34%) for an unobserved B_{1g} mode at 187 cm^{-1}, but not for any of the A_{1g} modes. Desbois et al. [100] have suggested an M—N(pyrrole) stretching assignment for a band near 250 cm^{-1} in (ImH)$_2$Fe^{II}EP, which shifts down 1 cm^{-1} upon pyrrole ^{15}N substitution; however, this band is probably the pyrrole out-of-plane tilting mode (see Sections B.iii.a. and C.i.b).

b. Out-of-Plane Modes. The same planar D_{4h} metalloporphyrin model, with point-mass C_b substituents and hydrogens at C_m, has 34 out-of-plane vibrational modes, classified as

$$\Gamma_{oop} = 3A_{1u} + 6A_{2u} + 4B_{2u} + 5B_{1u} + 8E_g.$$

Only the eight E_g modes are Raman-active, and their activation requires vibronic coupling between in- and out-of-plane electronic transitions, as discussed in Section B.iii.a. The A_{2u} modes are infrared-active. If, however, the mirror plane of the porphyrin is lost, and the symmetry is lowered to C_{4v} (e.g. via inequivalent axial ligands, or via distortion forces or electrostatic fields generated in a heme protein), then the A_{2u} modes transform as A_1, and Raman activity can thereby be induced in the infrared modes. Also $B_{1u} \rightarrow B_2(xy)$ and $B_{2u} \rightarrow B_1(x^2 - y^2)$, both of which are Raman-active (via coupling between in-plane electronic transitions, as with the B_{1g} and B_{2g} modes in D_{4h} symmetry), while $A_{1u} \rightarrow A_2(xy - yx)$, which is antisymmetric.

All of the out-of-plane modes are expected at relatively low frequencies, < 1000 cm^{-1}, since they involve no bond stretches, to a first approximation, but only deformations of the in-plane bonds. Warshel and Lappicirella [96] have used a semiempirical force field to calculate these modes for a metalloporphine; this has the same symmetry classification, but the heavy substituents are replaced by H, whose out-of-plane modes are at much higher frequencies. The results of this calculation are given in Table 3. The mode compositions are given in terms of the atom type around which deformation mainly occurs.

An alternative description is in terms of the local motions of the structural elements making up the porphyrin, i.e., the C_mH methine bridges, the pyrrole rings with their C_bS substituents, and the central metal atom. These motions are illustrated in Figure 11, while Table 4 classifies their contributions to the normal modes.

γ C_m—H modes have been identified in the i.r. spectrum (A_{2u}) of NiOEP [52] at 837 cm^{-1}, and as a depolarized band (E_g) at 841 cm^{-1} in the RR spectrum of (ImH)$_2$Fe^{II}PP [51]. Since the C_m—H bonds are well isolated, it is not surprising that the A_{2u} and E_g modes are nearly coincident. The B_{2u} mode no doubt also occurs at the same frequency.

Vibrational Assignments

Table 3. Out-of-Plane Modes of a Metalloporphine, Calculated with a Semiempirical Force Field (QCFF/P1)[a]

Sym.	Frequency (cm^{-1})	Description[b]	Sym.	Frequency (cm^{-1})	Description[b]
E_g	1064	χ_H			
B_{1u}	1063	χ_H	B_{1u}	414	χ_m
B_{2u}	1062	χ_H	E_g	373	χ_C
A_{2u}	855	χ_H	A_{1u}	290	χ_C
A_{2u}	852	$\chi_{H(m)}$	A_{2u}	278	χ_m
E_g	851	$\chi_{H(m)}$	E_g	172	χ_m
A_{2u}	788	χ_C	A_{2u}	153	χ_M
B_{1u}	788	χ_H	B_{2u}	146	χ_N
E_g	787	χ_H	E_g	130	χ_N
B_{1u}	594	χ_C	B_{1u}	77	
E_g	591	χ_C	B_{2u}	55	χ_C
A_{2u}	581	χ_C	A_{2u}	54	χ_M
A_{1u}	579	χ_C			
E_g	556	χ_C			
B_{2u}	541	χ_C			

[a] Reference [96]
[b] χ_y designates deformation around atom y = H (pyrrole H), H(m) (methine H), C (pyrrole C), m (methine C), N (pyrrole N), M (metal).

Five-membered aromatic heterocycles, such as pyrrole, have two out-of-plane folding modes, which are symmetric and antisymmetric with respect to the twofold axis (Figure 11). They are expected in the 400–600-cm^{-1} region [102]. Since there are four pyrrole rings in porphyrin, a total of eight folding modes are expected, classified as in Table 4; four of them occur as degenerate pairs (E_g). The calculation of Warshel and Lapicirella [96] shows a group of six modes in this region (Table 3). Ring-folding assignments are suggested for the [(ImH)$_2$Fe^{III}OEP]$^+$ RR bands at 403 and 469 cm^{-1} (see Figure 8). In this region no in-plane modes are calculated [90], nor are any observed in the NiOEP RR spectra [89]. Iron PP complexes [51] have analogous modes at $\simeq 425$ and $\simeq 500$ cm^{-1} (see Figure 6). The apparent frequency shifts are attributable to the altered peripheral substituents, and to activation of different pyrrole folding modes. In the case of Mb^{III}(F$^-$), the RR spectrum with B-excitation shows bands at 442, 472, and 501 cm^{-1} that are all shifted appreciably by deuteration at the C_α or C_β atoms of the vinyl group [51]. Such shifts can be understood on the basis of coupling between pyrrole folding and out-of-plane vinyl C—H(D) deformations. The altered frequencies, relative to protein-free complexes, are attributable to the influence of protein contacts on pyrrole mode activation.

As discussed in Section B.iii.a, ^{54}Fe-sensitive bands at 255 and 359 cm^{-1} are observed for [(ImH)$_2$Fe^{III}OEP]$^+$ but not [(CN)$_2$Fe^{III}OEP]$^-$, and are

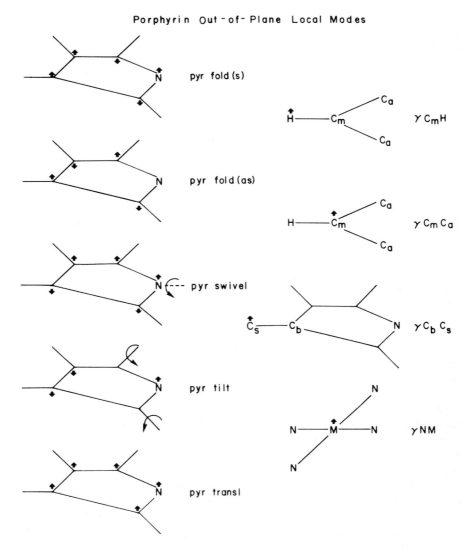

Figure 11 Illustration of local out-of-plane displacement coordinates, from which the porphyrin out-of-plane modes can be constructed, as shown in Table 4 (from [51]).

assignable to A_{2u}-type modes; they may involve pyrrole tilting and substituent deformation ($\gamma\ C_bS$). The C_mD-sensitive band at 320 cm^{-1} is assignable to methine deformation, $\gamma\ C_aC_m$ [51]. The same A_{2u}-activation mechanism would fail to bring out pyrrole swiveling or translation modes, since these local motions do not contribute to A_{2u}-modes. They probably occur at quite low frequencies. Warshel and Lapicirella [96] calculate several out-of-plane modes

Vibrational Assignments

Table 4. Local-Mode Contributions to the Out-of-Plane Normal Modes of a D_{4h} OEP-Type Metalloporphyrin, with Point Mass Pyrrole Substituents S

Local-Mode[a]	A_{1u}	A_{2u}	B_{1u}	B_{2u}	E_g
$\gamma C_m H$		1	1		1
pyr fold (s)		1		1	1
pyr fold (as)	1		1		1
$\gamma C_b S$	1	1	1	1	2
$\gamma C_a C_m$		1	1		1
pyr tilt		1		1	1
pyr swivel	1		1		1
pyr transl.		(1)[b]		1	(1)[c]
γNM		1			
Out-of-plane vibrational representation	3	6	5	4	8

[a] See Figure 11 for descriptions.
[b] Molecular translation (z).
[c] Molecular rotations (R_x, R_y).

below 200 cm^{-1} (Table 3). Two of these, 153 and 54 cm^{-1}, are calculated to have major γ NM contributions. For metalloporphyrins with axial ligands the predominantly γ NM mode should be at a very low frequency, since it involves a concerted out-of-plane displacement of the entire metal-ligand group.

ii. Peripheral Substituent Effects

As long as the peripheral substituents are connected to the pyrrole rings via saturated carbon atoms, the RR pattern is expected to be quite similar to that of OEP. Electronic coupling is insignificant, and mechanical coupling is largely limited to interaction with the C_b—S stretching and bending vibrations. It is likely however that coupling with substituent C—H bending vibrations is important in the $\simeq$ 1100-cm^{-1} region, which is quite complicated for OEP [89]. Some differences in this region are seen for analogous complexes of OEP and of MP (which has the same substituent pattern as PP, but with ethyl instead of vinyl groups), although the spectra are otherwise very similar [103].

a. Vinyl Modes of Protoporphyrin IX. Protoporphyrin IX, which occurs in Hb, Mb, peroxidase, catalase, cytochrome P450, and b-type cytochromes, has two peripheral vinyl groups (Figure 1) which are capable of conjugation with the porphyrin π-system. The $\simeq$ 10-nm red shift of the B and Q transitions in PP, relative to MP or OEP complexes, is evidence for such conjugation, at least in the excited state. The internal modes of the vinyl groups are subject to resonance enhancement, via their involvement in the excited-state distortion of the extended chromophore and via mechanical coupling with porphyrin skeletal modes. Aside from the C—H stretching vibrations, which are not expected to be resonance-enhanced, there are seven internal modes of the —CH=CH$_2$

group, including the C=C stretch (ν C=C), three in-plane C—H bends (δ C—H) and three out-of-plane C—H bends (γ C—H). In addition there are modes involving C_b—vinyl stretching and C_b—C_α—C_β and C_b—C_b—C_α bending.

These modes are illustrated in Figure 12. All of them, except the lowest-frequency out-of-plane C—H bend and δ $C_bC_bC_\alpha$, have been located in RR and IR spectra of NiPP, via the frequency shifts produced by deuterium substitution at C_α or C_β [104] (see Figure 13). In some cases, selective coupling was found between skeletal modes and the members of the pairs of vinyl modes representing in- and out-of-phase combinations of the local modes, producing splittings of these pairs by up to 100 cm^{-1}.

In addition the asymmetric disposition of the vinyl groups on adjacent pyrrole rings (Figure 1), destroys the inversion center of the chromophore, and induces Raman activity into the infrared modes (E_u in D_{4h} symmetry). Several of these modes have been identified in the NiPP RR spectra, via correlations with the NiOEP E_u modes [104].

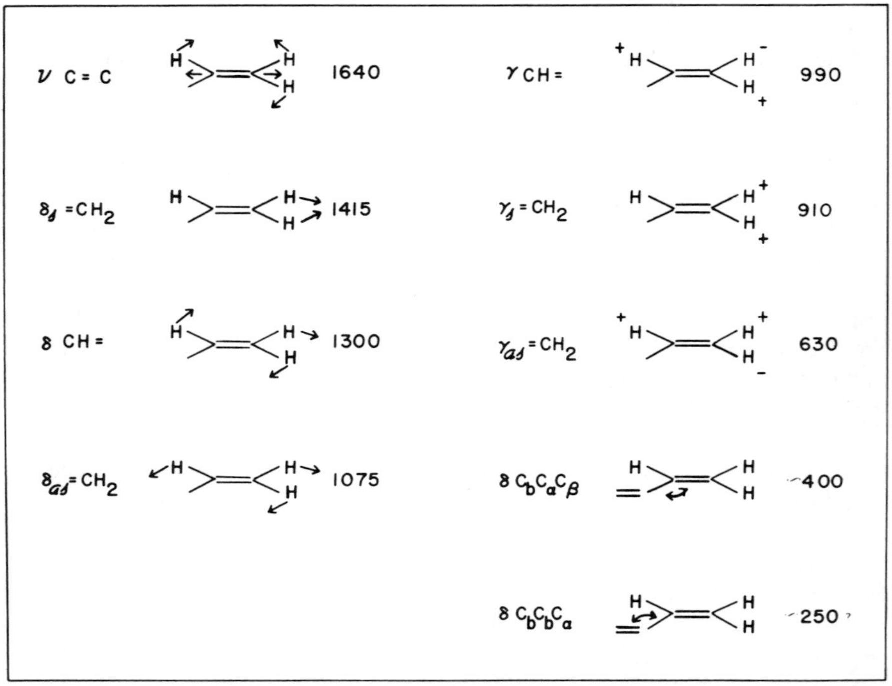

Figure 12 Schematic diagram of the vinyl modes. Typical frequencies (cm^{-1}) are indicated. (From [104].)

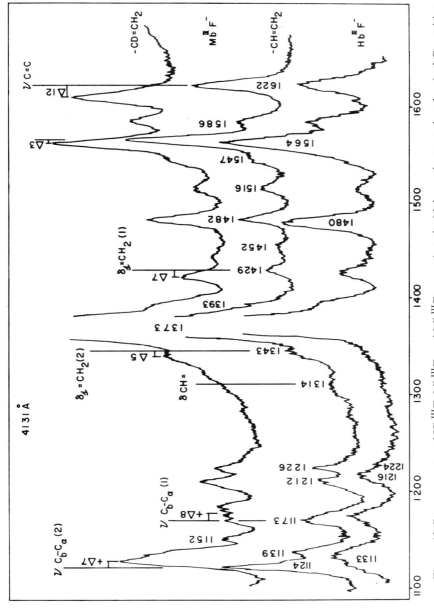

Figure 13 Raman spectra of $Hb^{III}F$, $Mb^{III}F$, and $Mb^{III}F$ reconstituted with heme deuterated at the vinyl C_α position, with 4131-Å Kr^+ excitation. Note the d_α shifts of several vinyl modes, including $\nu C_b-C_\alpha(2)$, which is activated by the protein. (From [103].)

Assignments of all the observed NiPP modes are given in Figure 14. The vinyl and E_u-type modes complicate the RR spectra considerably. For example, the region between 1450 and 1650 cm^{-1}, which is sensitive to the porphyrin structure (see below), contains no fewer than nine bands. These can overlap appreciably, but it is possible to sort them out with polarization measurements and variable-wavelength excitation [104].

The coupling between vinyl and porphyrin skeletal modes produces some frequency shifts of the latter. In particular, the highest-frequency A_{1g} mode, which couples with the nearby C=C stretch, is 10 cm^{-1} lower in NiPP than in NiOEP (1592 ← 1602 cm^{-1}) and is shifted out of coincidence with the highest-frequency A_{2g} mode (1602 cm^{-1}). Also, the B_{2g} and A_{2g} modes at 1409 and 1396 cm^{-1} in NiOEP are shifted down and up, respectively, to 1401 and 1399 cm^{-1} via coupling with the =CH$_2$ scissors modes. The $\simeq$ 1100-cm^{-1} region, where C_b—S (including C_b—vinyl) contributions dominate, shows several differences between NiPP and NiOEP.

b. Heme a. The terminal oxidase of mitochondria, cytochrome oxidase, contains two copper ions and two molecules of heme a [105]. One of the latter, heme a_3, is accessible to exogenous ligands, which can switch the iron ion from a high- to a low-spin state. The heme a_3, along with one of the copper ions, to which it appears to be antiferromagnetically coupled in the oxidized form, is believed to be the site of O$_2$ binding. The nonaccessible heme a remains low-spin in all states of the protein, and its role is probably limited to electron transfer. Several RR studies of cytochrome oxidase and of isolated heme a complexes have been reported [106–115]. The protein spectra are complicated by the coexistence of the two heme a molecules. In those forms containing high-spin heme a_3, however, the spectral contributions of heme a and heme a_3 can to some extent be disentangled by selective laser excitation, since the absorption wavelengths differ appreciably [112, 113]. The protein spectra are frequently subject to interference from fluorescence and photoreduction [108, 109], but these problems can be minimized by careful purification and by judicious choice of laser wavelength [112, 113]. Woodruff et al. [114] have suggested that, in addition to heme a modes, vibrations due to copper may be observable in the low-frequency region.

Heme a differs from protoheme (Figure 1) in having a vinyl and a formyl substituent attached to opposite pyrroles, rather than two vinyl substituents on adjacent pyrroles. The formyl group is strongly electron-withdrawing, and shifts the porphyrin π-π^* transitions to the red. The addition of bisulfite or cyanide to the formyl group results in both absorption and RR spectra resembling those of protoheme [107].

The formyl νC=O stretch was earlier identified [106] as the highest-frequency band, $\simeq$ 1670 cm^{-1}, in the cytochrome oxidase spectrum. In contrast to νC=C of protoheme [103], νC=O of heme a is somewhat dependent on the iron oxidation state [114] and on the spin state [116], and it is also

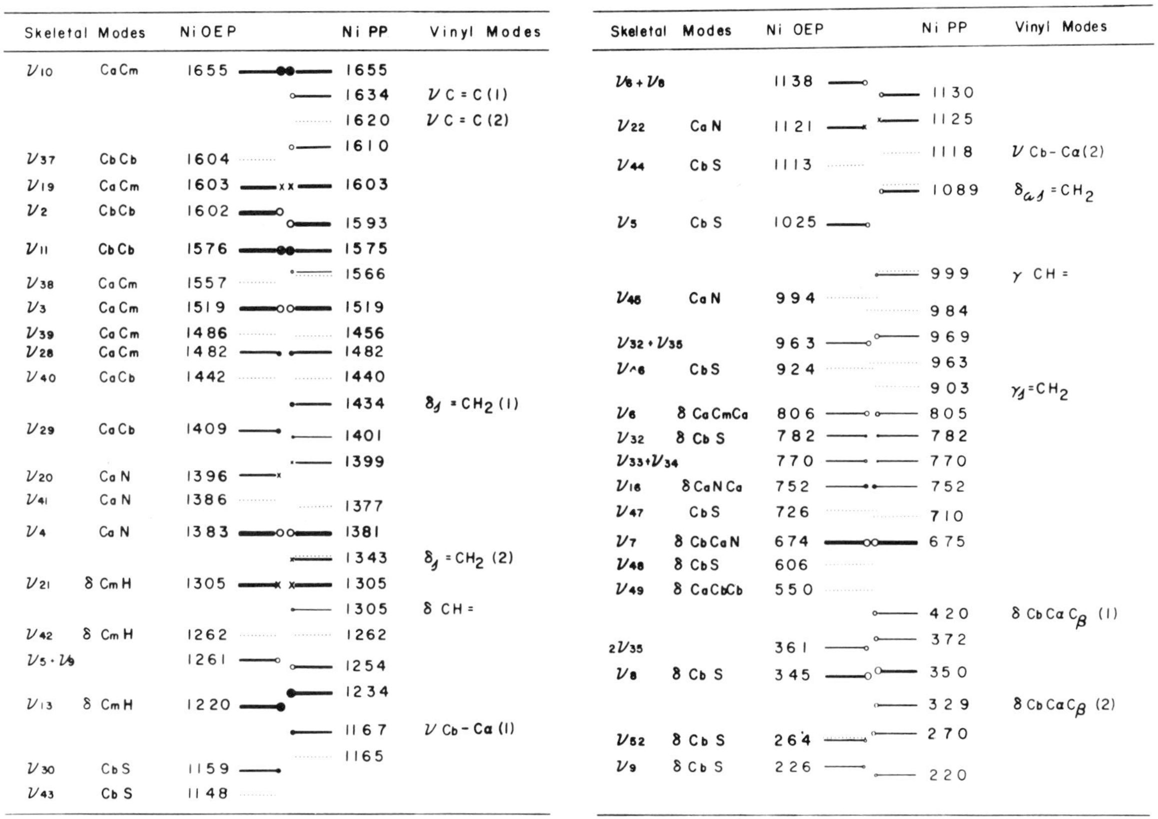

Vibrational Assignments

Skeletal Modes		Ni OEP	Ni PP	Vinyl Modes		Skeletal Modes		Ni OEP	Ni PP	Vinyl Modes
ν_{10}	CaCm	1655	1655			$\nu_6 + \nu_8$		1138	1130	ν Cb-Ca(2)
			1634	ν C=C(1)					1125	
			1620	ν C=C(2)		ν_{22}	CaN	1121	1118	δ_{as}=CH$_2$
ν_{37}	CbCb	1604	1610			ν_{44}	CbS	1113	1089	
ν_{19}	CaCm	1603	1603			ν_5	CbS	1025		γ CH =
ν_2	CbCb	1602	1593							
ν_{11}	CbCb	1576	1575							γ_j=CH$_2$
ν_{38}	CaCm	1557	1566			ν_{46}	CaN	994	999	
ν_3	CaCm	1519	1519						984	
ν_{39}	CaCm	1486	1456			$\nu_{32} + \nu_{35}$		963	969	
ν_{28}	CaCm	1482	1482			ν_{48}	CbS	924	963	
ν_{40}	CaCb	1442	1440						903	
			1434	δ_j=CH$_2$(1)		ν_6	δ CaCmCa	806	805	
ν_{29}	CaCb	1409	1401			ν_{32}	δ CbS	782	782	
			1399			ν_{33}, ν_{34}		770	770	
ν_{20}	CaN	1396	1377			ν_{16}	δCaNCa	752	752	
ν_{41}	CaN	1386	1381			ν_{47}	CbS	726	710	
ν_4	CaN	1383				ν_7	δ CbCaN	674	675	
			1343	δ_j=CH$_2$(2)		ν_{48}	δ CbS	606		
ν_{21}	δ CmH	1305	1305	δ CH=		ν_{49}	δ CaCbCb	550		
ν_{42}	δ CmH	1262	1262						420	δ CbCaCβ(1)
$\nu_5 \cdot \nu_6$		1261	1254			$2\nu_{35}$		361	372	
ν_{13}	δ CmH	1220	1234			ν_8	δ CbS	345	350	
			1167	ν Cb-Ca(I)					329	δ CbCaCβ(2)
			1165			ν_{52}	δ CbS	264	270	
ν_{30}	CbS	1159				ν_9	δ CbS	226	220	
ν_{43}	CbS	1148								

Figure 14 Schematic diagram of the RR (solid lines) and i.r. (dashed lines) modes of NiOEP and NiPP, with assignments of the skeletal (left) and vinyl (right) modes. The major contributor to each skeletal mode is indicated next to the mode number. The symbols indicate the RR polarizations: ○, polarized; ●, depolarized; ×, anomalously polarized. The sizes of the symbols indicate the relative intensities of the RR bands at the wavelengths of greatest enhancements. (From [104].)

affected by H-bonding [115, 116]. As with νC=C of protoheme, both νC=C and νC=O of heme a are observable with B- but not Q-band excitation [116]. Other formyl modes have been identified via H-D exchange at the formyl group [116]. In benzaldehyde [117], the in-plane H—CO bend occurs at 1389 cm^{-1} and shifts down to 1043 cm^{-1} upon deuteration, while the ϕ—(CHO) stretch is found at 1206 cm^{-1}, and shifts *up* to 1217 cm^{-1} upon deuteration, due to interaction with the H(D)—CO bend. In [(ImH)$_2$Fe^{III}Pa]$^+$ (Pa = porphyrin a) a RR band at 1220 cm^{-1} shifts up to 1230 cm^{-1} upon deuteration, and is assigned to the C$_b$—CHO stretch. The region of the H—CO bend is obscured by a skeletal mode at 1390 cm^{-1}, but an apparent 2-cm^{-1} upshift of this RR band upon deuteration gives evidence of δH—CO mode contribution on the low-frequency side. Neither the out-of-plane, γH—CO bend [1009 (863) cm^{-1} in benzaldehyde ($-d_1$)] nor the δCO bend [649 (641) cm^{-1} in benzaldehyde ($-d_1$)] has been observed in [(ImH)$_2$Fe^{III}Pa]$^+$.

The skeletal modes of heme a have been catalogued from RR and i.r. spectra [116]. In Table 5, they are compared with the protoheme frequencies for the (ImH)$_2$Fe^{II} complexes. There are a number of small frequency dif-

Table 5. Comparison of Skeletal and Peripheral Mode Frequencies (cm^{-1}) for (ImH)$_2$Fe^{II}P (P = Porphyrin a versus Protoporphyrin)

Assignment[a]	Pa[b]	PP[b]	Assignment	Pa	PP
νC=O	1644		γCH=		1008
ν_{10}(C$_a$C$_m$)	1620	1617	ν_{45}(C$_a$N)	995	995
νC=C	1622	1620	ν_{46}(C$_b$S)	933	925
ν_{37}(C$_b$C$_b$)	{1614, 1583}	1604	γC$_m$H	841	841
ν_2(C$_b$C$_b$)	1593	1584	ν_6(δC$_a$C$_m$C$_a$)		819
ν_{19}(C$_a$C$_m$)	1588	1583	ν_{32}(δC$_b$S)	792	791
ν_{11}(C$_b$C$_b$)	1504	1539	ν_{16}(δC$_a$NC$_a$)	746	748
ν_{38}(C$_a$C$_m$)	{1574, 1546}	1560	ν_{47}(δC$_b$S)	714	717
ν_3(C$_a$C$_m$)	1493	1493	ν_7(δC$_b$C$_a$N)	676	675
ν_{28}(C$_a$C$_m$)	1469	1461			
δ=CH$_2$(1)		1431			
ν_{29}(C$_a$C$_b$)	1390	1390			
ν_{20}(C$_a$N)	1392	1392			
ν_4(C$_a$N)	1359	1359			
δ=CH$_2$(2)	1332	1337			
ν_{21}(δC$_m$H)	1305	1306			
δCH=	1308	1306			
νC$_b$—CHO	1220				
ν_{13}(δC$_m$H)	1226	1225			
νC$_b$—C$_\alpha$(1)	1168	1174			
ν_{22}(C$_a$N)	1128	1125			
δ_{as}=CH$_2$	1084	1089			

[a] Skeletal frequency mode numbers as in Table 2, with major potential-energy contributor in parentheses.
[b] Pa, porphyrin a; PP, protoporphyrin IX.

ferences associated with the change in peripheral substitution. Both high frequency E_u modes, ν_{39} and ν_{38}, are split, by ~ 30 cm^{-1}, and the two components are selectively enhanced at different wavelengths in the RR spectra. The E_u splitting and the selective enhancement are attributable to the strong x, y inequivalence induced by the opposed formyl and vinyl groups.

c. *Tetraphenylporphines.* While substituent alterations among the physiological-type porphyrins influence the vibrational spectrum in various subtle ways, the meso-tetraphenylporphines present a substantially altered vibrational pattern. These synthetic porphyrins, whose ease of preparation and of crystallization has made them quite useful in structural studies of porphyrin chemistry, have hydrogen atoms attached to the pyrrole C_b atoms, and phenyl rings attached to the methine bridges (Figure 1). Vibrational couplings are thereby altered substantially. For example the δ C_b—H bending coordinates mix into the high-frequency modes of all symmetries [35], while in OEP-related porphyrins, the δ C_m—H coordinates do not contribute to the A_{1g} or B_{1g} modes.

Moreover, the RR spectra contain bands associated with modes of the phenyl ring (1600, 1030, 995, and 890 cm^{-1} [35]). These have been identified via their shifts upon phenyl perdeuteration [35]. RR enhancement of phenyl modes has been suggested to indicate significant phenyl conjugation in the porphyrin π-system [118], but conjugation in the ground state cannot be very important in view of the large phenyl tilt angles and barriers to rotation [119] and the small NMR contact shifts of the phenyl protons [120]. Phenyl interaction in the porphyrin π-π* excited state is sufficient to explain phenyl mode enhancements. It has been suggested [35] that since the Q-transition transfers electron density to the C_m atoms, there is a driving force for rotating the phenyl rings into the porphyrin plane to allow delocalization via the phenyl π-system. This might account for the large origin shifts for certain modes involving the pyrrole rings, evidenced by 0-0, 0-1, and 0-2 maximization in the excitation profiles [35], which are not seen for OEP-related porphyrins. These modes could provide distortion pathways to relieve the steric hindrance to phenyl rotation [35].

Despite the differences between TPP- and OEP-related porphyrins, the pattern of porphyrin skeletal vibrations remains recognizable, as shown by the comparison between [(ImH)$_2$Fe^{III}PP]$^+$ and [(ImH)$_2$Fe^{III}TPP]$^+$ skeletal frequencies shown in Table 6. There is a good correlation in the frequency ordering between modes of given symmetries, although the frequencies themselves differ by as much as 60 cm^{-1}. The mode compositions, however, may differ by much more than the correlation suggests. For example, the deuteration shifts [35] show that the A_{1g} mode correlating with $\nu_3(C_aC_m)$ of PP is mainly C_aC_b in character. The altered compositions are attributable to the very different couplings of the bond stretches when the heavy substituents switch places with the hydrogen atoms at the C_m and C_b positions.

Table 6. Suggested Correlations Between Porphyrin Skeletal Modes of $[(ImH)_2Fe^{III}PP]^+$ and $[(ImH)_2Fe^{III}TPP]^+$

			Frequency (cm^{-1})	
Mode [a]	Sym.	Type [b]	$[(ImH)_2Fe^{III}PP]^+$ [104]	$[(ImH)_2Fe^{III}TPP]^+$ [35]
ν_{10}	B_{1g}	C_aC_m	1640	1582
ν_{19}	A_{2g}	C_aC_m	1586	1568
ν_2	A_{1g}	C_bC_b	1579	1540
ν_{11}	B_{1g}	C_bC_b	1562	1505
ν_3	A_{1g}	C_aC_m	1502	1456
ν_4	A_{1g}	C_aN	1373	1370

[a] Skeletal mode number for NiOEP (Table 2).
[b] Major potential energy contributor for NiOEP.

iii. Axial-Ligand Modes

a. M—L Stretching

M^{II}—XY. Table 7 lists the frequencies of the metal–axial-ligand stretching modes that have been assigned, mostly on the basis of ligand or metal isotope frequency shifts. For the diatomic ligands that bind to deoxyHb and deoxyMb, the Fe—XY stretching frequencies increase in the order CO < NO < O_2. This is also the order of decreasing π^*-orbital energies, and the stretching frequencies reflect the expected trend of Fe d_π back donation to these π-acceptor ligands. The same trend is seen in the π-sensitive porphyrin skeletal frequencies, which reflect the competition between porphyrin and axial-ligand π^* orbitals for the d_π electrons [43, 60, 61] (see Section D.ii). When the *trans* axial ligand of NO-heme is lost, ν Fe—NO increases by the amount expected on the basis of the decreased Fe—NO distance observed in TPP crystal structures [130], which is associated with the NO *trans* effect.

The high Fe—O_2 stretching frequency, $\simeq 570$ cm^{-1}, bears on the long-standing controversy [131–135] over the electronic structure of oxy heme. While this complex is formed from O_2 and Fe^{II} porphyrin, many spectroscopic properties, including the high-frequency RR spectrum [136, 43] are more consistent with low-spin Fe^{III} than Fe^{II}, and the O—O stretching frequency, observed by infrared spectroscopy [137], is in accord with an Fe^{III}—O_2^- formulation. The transfer of a d_π electron to O_2 appears to be essentially complete, but, of course, this transfer is compensated by σ-donation from O_2 to Fe; electronic-structure calculations [138] indicate that the *net* transfer of charge is small. The extent of back donation depends sensitively on the match of d_π and π^* orbital energies, and may be affected, for example, by alterations in the *trans* axial ligand [135]. The interaction is expected to be less for Co^{II} than for Fe^{II}, and the lower νCo—O_2 observed for CoHbO_2 and CoMbO_2 [87] supports this expectation. The O—O frequency appears to be less sensitive [87, 135] to the difference in electronic structure than is the M—O_2 frequency.

Table 7. Assigned Metal–Axial-Ligand Stretching Frequencies of Metalloporphyrins

Mode	Frequency (cm^{-1})	Isotope Shift (cm^{-1}) [a]	Molecule	Reference
		(a) Imidazole—M^{II}—XY species (low-spin)		
Fe—O$_2$	567	27 ^{18}O$_2$	HbO$_2$	[63]
	561–572		O$_2$(N—MeIm)Fe^{II}TP$_{iv}$PP	[121–123]
	572		MbO$_2$	[124]
Co—O$_2$	537	23 ^{18}O$_2$	(Co)HbO$_2$	[87]
	539		(Co)MbO$_2$	[87]
Fe—NO	549	10 ^{15}NO	HbNO	[67]
	553	7 ^{15}NO	HbNO	[62] [d]
	553, 592	7, 3 ^{15}NO	HbNO + IHP	[62] [d]
Fe—CO	505	13 ^{13}C^{18}O	LegHbCO	[68] [e]
	507	4 ^{13}C, 9 ^{18}O, 13 ^{13}C^{18}O	HbCO	[68] [f]
	512	3 ^{13}C, 8 ^{18}O, 15 ^{13}C^{18}O	MbCO	[68] [f]
Fe—ImH	271	3 ^{18}O$_2$	MbO$_2$	[59]
		(b) Imidazole—Fe^{III}—X$^-$ species		
Fe—N$_3^-$	411	6 ^{15}N$_3^-$	Mb^{III}(N$_3^-$)	[81]
	413		Hb^{III}(N$_3^-$)	[78, 81]
Fe—OH$^-$	497	20 ^{18}OH$^-$	Hb^{III}(OH$^-$)	[78]
	490	22 ^{18}OH$^-$	Mb^{III}(OH$^-$)	[79]
	490	19 ^{18}OH$^-$, 2 ^{54}Fe	Mb^{III}(OH$^-$)	[80]
Fe—F$^-$	462	2 ^{54}Fe	Mb^{III}(F$^-$)	[80]
	461, 422		Mb^{III}(F$^-$)	[79]
	471, 443		Hb^{III}(F$^-$)	[78]
Fe—ImH	248	3 ^{54}Fe	Mb^{III}(H$_2$O)	[129 (1981)]
	267	3 ^{54}Fe	HRP^{III}(F$^-$)	[129 (1981)]
		(c) M^{III}—X$^-$ (5-coordinated)		
Fe—S(cys)	351	4.6 ^{34}S, 2.4 ^{54}Fe	cytP450(Fe^{III} [c])	[126]
Fe—F$^-$	606	3 ^{54}Fe	(F)Fe^{III}OEP	[101, 73, 74]
Fe—Cl$^-$	360		(Cl)Fe^{III}OEP	[101, 74]
Fe—Br$^-$	270		(Br)Fe^{III}OEP	[101, 74]
Fe—I$^-$	246		(I)Fe^{III}OEP	[101]
Fe—NCS$^-$	315	1.5 ^{54}Fe	(NCS)Fe^{III}OEP	[101]
Fe—N$_3^-$	421	3 ^{54}Fe	(N$_3$)Fe^{III}OEP	[101]
Mn—F$^-$	495		(F)Mn^{III}EP	[55]
Mn—Cl$^-$	285	4 $^{37/35}$Cl	(Cl)Mn^{III}EP	[55]
Mn—Br$^-$	245		(Br)Mn^{III}EP	[55]
Mn—I$^-$	233		(I)Mn^{III}EP	[55]
		(d) L—Fe—L		
Fe^{III}—(ImH)$_2 \nu_s$	200	4 ImH—d_3	[(ImH)$_2$Fe^{III}PP]$^+$	[69] [g]
ν_{as}	377	3.5 ^{54}Fe	[(ImH)$_2$Fe^{III}OEP]$^+$	[101]
Fe^{II}—(ImH)$_2 \nu_s$	200	4 ImH—d_3, 2ImH—^{15}N	(ImH)$_2$Fe^{II}PP	[69]
Fe^{II}—(py)$_2 \nu_s$	179	5 py—d_5	(py)$_2$Fe^{II}PP	[84]
		(e) Fe—Imidazole (5-coordinated)		
	220	2 (2-MeImH—d_5)	(2-MeImH)Fe^{II}PP in H$_2$O	[70]
	207	3 ^{54}Fe	(2-MeImH)Fe^{II}PP in CTAB [b]	[71, 123]
	205		(2-MeImH)Fe^{II}OEP in C$_6$H$_6$	[127]
	195		(1,2-Me$_2$Im)Fe^{II}PP in H$_2$O	[127]
	195		(1,2-Me$_2$Im)Fe^{II}OEP in C$_6$H$_6$	[127]
	209	3 (2-MeImH—d_5), 3 ^{54}Fe	(2-MeImH)Fe^{II}TP$_{iv}$PP in CH$_2$Cl$_2$	[123]

Table 7. Assigned Metal–Axial-Ligand Stretching Frequencies of Metalloporphyrins

Mode	Frequency (cm^{-1})	Isotope Shift (cm^{-1})[a]	Molecule	Reference
	225		(1-MeIm)Fe^{II}TP$_{iv}$PP in CH$_2$Cl$_2$	[123]
	200		(1,2-Me$_2$Im)Fe^{II}TP$_{iv}$PP in CH$_2$Cl$_2$	[123]
	239		(2-MeIm$^-$)Fe^{II}PP in DMF	[127]
	207, 223		deoxyHb	[70, 128, 183]
	220		deoxyMb	[71]
	244		HRP	[129]

[a] Down- or upshift upon substitution of the indicated isotope.
[b] CTAB = cetyltinethylammorium bronide detergent in H$_2$O.
[c] Substrate-bound.
[d] Stong et al.
[e] Armstrong et al.
[f] Tsubaki et al.
[g] Mitchell et al.

M^{III}—X^-. Stretching modes for anionic ligands have been identified for N_3^-, OH$^-$, and F$^-$ complexes of Hb^{III} and Mb^{III} [Table 7(b)]. The lowest frequency, 413 cm^{-1}, is observed for N_3^-, although it has been assigned, on the basis of its temperature dependence, to the low-spin component of the azide spin mixture [81]. For OH$^-$, the band appears to be due to the high-spin component, as judged by its excitation profile [78]. Its frequency, 497 cm^{-1}, is, however, much higher than that of N_3^-, although one would expect the M—L force constant to be higher for low-spin than for high-spin complexes (the effective mass of N_3^-, however, is uncertain). Two bands, 471 and 443 cm^{-1}, have been assigned to Fe—F stretching in Hb^{III}F$^-$, based on their excitation profiles [78]; the lower frequency was suggested to be due to H-bonding to a water molecule having partial occupancy of a site in the heme pocket [78]. Both frequencies are appreciably lower in Mb: 461 and 422 cm^{-1}. The higher-frequency band has been shown [80] to have a 2-cm^{-1} ^{54}Fe isotope shift, confirming the Fe—F assignment.

The Fe—F frequency seen for 5-coordinated heme fluoride is much higher, 606 cm^{-1} [Table 7(c)], because the absence of a *trans* axial ligand allows the Fe^{3+} ion to move out of the heme plane and form a strong bond to F$^-$ [78]. The 5-coordinated N_3^- frequency, 421 cm^{-1}, is only slightly higher than that seen for Hb^{III}N$_3^-$, reflecting the compensating effects of the absence of a *trans* ligand, and the high → low-spin transition. The recently assigned [126] Fe—S(cys) stretch, 351 cm^{-1}, of substrate-bound cytochrome P450$_{cam}$ is at nearly the same frequency as the Fe—Cl stretch in (Cl$^-$)Fe^{III}OEP [102], 360 cm^{-1}, confirming five-coordination of the heme in this form of the protein [126]. Curiously, the M^{III}—halide frequencies are much lower for Mn^{III} than for Fe^{III} five-coordinated porphyrins [Table 7(c)]. Kitagawa et al. [74] have questioned the Mn—Cl assignment, 285 cm^{-1}, as being too low, but the 4-cm^{-1} $^{37/35}$Cl$^-$ isotope shift [55] is compelling evidence. It appears that the Mn^{III}—halide bonds are much weaker than the Fe^{III}—halide bonds.

Fe—Imidazole. Because the heme iron atom is bound to an imidazole group in most hemeproteins, the Fe—ImH stretching frequency is of particular importance, but it has also proved to be elusive. Because the effective mass of the rigid imidazole ring is high, Fe—ImH modes are expected to fall in the 200–300-cm^{-1} range, where low-frequency porphyrin modes interfere. Metal-imidazole modes are generally found in this region [139].

For the bis-pyridine complex (py)$_2$Fe^{II}MP, the symmetric py—Fe—py stretch was located at 179 cm^{-1} via the 5-cm^{-1} downshift upon ligand perdeuteration [Table 7(d)]. A normal-coordinate analysis showed this mode to be a breathing motion of essentially rigid py rings [84]. Indeed, the frequency was well approximated with a linear triatom calculation, using the same Fe—py force constant, in which the py "atom" was given the full mass of the ring. The mass of the ImH ring is only slightly less than that of py, and the symmetric ImH—Fe—ImH mode has now been located [Table 7(e)] at a slightly higher frequency, 200 cm^{-1}, for (ImH)$_2$Fe^{II}PP via ImH ^{15}N [69] (Desbois and Lutz) and $-d_3$ [69] (Mitchell et al.) isotope shifts.

Surprisingly, it is found at the same frequency for [(ImH)$_2$Fe^{III}PP]$^+$ [69], (Mitchell et al.) although the increase in oxidation state would have been expected to increase the bond strength, and therefore the stretching frequency. Apparently the extra electron in (ImH)$_2$Fe^{II}PP is completely delocalized to the porphyrin ring, as reflected in the lowered porphyrin skeletal frequencies [61] (see Section D.ii), leaving the Fe—ImH bond strength unaltered. The ImH—Fe—ImH asymmetric stretch of [(ImH)$_2$Fe^{III}OEP]$^+$ has been identified at 377 cm^{-1} in the i.r. spectrum, via its ^{54}Fe shift [101]. The large frequency difference between the in- and out-of-phase stretches, which depends largely on the ligand-metal mass ratio, is further evidence of the high effective mass of the ImH ligand. A triatom calculation with the full ImH mass gives a reasonable Fe—ImH force constant, $K = 1.63$ md/Å [69] (Mitchell et al.).

When one of the ImH ligands is replaced by a ligand, L, of smaller mass, there will be two Fe—ligand modes, both Raman- and i.r.-active. The higher-frequency mode will be primarily Fe—L stretching, while the lower one will be primarily Fe—ImH stretching. The two modes will remain coupled, however, to an extent that depends on the ligand mass (and force-constant) disparity. This was the basis of a search for the Fe—ImH mode in MbO$_2$, using the expected ^{18}O$_2$ frequency shift [58]. A linear triatom calculation, using the Fe—ImH force constant obtained from [(ImH)$_2$Fe^{III}PP]$^+$, and the known frequency of the Fe—O$_2$ mode, gave a predicted Fe—ImH frequency in the 247–263-cm^{-1} range, depending on the effective mass assumed for O$_2$, and an ^{18}O$_2$ shift of 3–4 cm^{-1}. The MbO$_2$ RR spectrum showed a $\simeq$ 3-cm^{-1} shift in a weak shoulder at 271 cm^{-1}, and this was tentatively assigned to the Fe—ImH mode [Table 7(a)]. For HbO$_2$ this region is more heavily overlapped by porphyrin modes and no definite assignment could be obtained [58].

For five-coordinated high-spin Fe^{II} porphyrins prepared with the sterically hindered imidazole, 2-MeImH [140], the Fe—ImH frequency has been located at $\simeq$ 220 cm^{-1} via its shift on ligand perdeuteration [70] or on ^{54}Fe substitution [71, 123]. Desbois et al. [80] have instead assigned this band to

Fe—pyrrole stretching, and indeed, its ^{15}N—pyrrole shift, 1.5 cm^{-1} [100], does imply appreciable involvement of the pyrrole N atoms in the mode. Nevertheless the frequency shifts on 2-MeImH perdeuteration (Desbois et al. [100] suggest that this might be due to the steric effect of the 2-CD$_3$ group, but D is less sterically demanding than H [141], and this effect should have led to an increase, not a decrease, in the frequency), and ^{54}Fe substitution clearly demonstrate the predominant contribution of Fe—N(2-MeImH) stretching. The ^{54}Fe shift implies movement of the Fe atom perpendicular to the porphyrin plane, i.e., along the Fe—N(2-MeImH) bond. Movement parallel to the plane cannot occur for Raman modes, due to the symmetry of the complex in the plane.

A diatomic calculation, using the full 2-MeImH mass, gave a force constant of 0.96 md/Å [70]. This is substantially smaller than that obtained for $[(ImH)_2Fe^{III}PP]^+$, as expected, since the Fe^{II} ion is high-spin and the Fe—N(2-MeImH) bond is lengthened [72]. The reason that the Fe—ImH mode nevertheless appears at a higher frequency for (2-MeIm)Fe^{II}PP is that the reduced mass for a diatom Fe—L, namely $(1/m_{Fe} + 1/m_L)^{-1}$, is less than the reduced mass for the symmetric mode of a linear triatom, L—Fe—L, which is just m_L [139].

The Fe—N(2-MeImH) frequency is lowered appreciably, to $\simeq 205$ cm^{-1}, when the complex is dispersed in a detergent [71, 123] or when it is dissolved in benzene [127]. The increased frequency in aqueous solution has been attributed [127] to an increase in Fe—N(2-MeImH) bond strength due to H-bonding of the N$_1$ proton to water molecules, which would enhance the donor properties of 2-MeImH. When 1,2-diMeIm, in which N$_1$ is methylated and no H-bonding is possible, was used as the imidazole, the Fe—N(1,2-diMeIm) frequency was the same, 195 cm^{-1}, in water and in benzene [127]. When the 2-ImH proton was removed completely, with potassium t-butoxide in dimethylformamide, the Fe—N(2-MeIm$^-$) frequency was upshifted to 239 cm^{-1} [127].

In deoxyHb [70, 128] and deoxyMb [71], for which the 2-MeImH adducts are models, the Fe—ImH frequency is close to 220 cm^{-1}, presumably reflecting the H-bond of the N$_\varepsilon$ proton to a backbone carbonyl group [142]. In reduced (high-spin Fe^{II}) horseradish peroxidase (HRP), however, the mode is found at a much higher frequency, 244 cm^{-1} [129], similar to that resulting from 2-MeImH deprotonation [127]. This inference is supported by other spectroscopic evidence [143] that the proximal imidazole is deprotonated, or strongly H-bonded, in reduced HRP.

The (high-spin) Fe^{III}—ImH frequency has been located [129 (1981)] by ^{54}Fe substitution at 248 cm^{-1} in Mb^{III}(H$_2$O) and 267 cm^{-1} in HRP^{III}(F$^-$), while a 274-cm^{-1} band in native HRP^{III} was likewise assigned to this mode. The increase between Mb^{III}(H$_2$O) and HRP^{III}(F$^-$) was suggested to be due to strong H-bonding of the proximal imidazole, as in reduced HRP, while the additional increase in native HRP was suggested to arise from the absence of a sixth ligand (see Section E.ii).

b. Internal Ligand Modes. As noted in Section B.iii.b, internal modes of the axial ligands can be enhanced via L → M or M → L CT transitions. The first example of this effect in heme RR spectra was the observation of enhanced pyridine modes for bis-pyridine hemochromes [61]. Their excitation profiles were distinct from those of the porphyrin modes [84], and coincided with a bump on the absorption spectrum at ≃ 490 nm, which had been suggested [85] to arise from an Fe^{II} → py CT transition. This assignment was supported by an intensity analysis [84] of the py modes of $(py)_2Fe^{II}MP$, and of its py-d_5 analog, which showed the excited-state geometry to be consistent with population of the first py π^*-orbital.

Azide has relatively high-lying filled orbitals, and its CT transitions to Mn^{III} [55] and Fe^{III} [81] in Mb adducts have been shown to enhance both stretching and bending vibrations of the N_3^- ligand. Likewise the O—O stretch of bound dioxygen is enhanced [87] by what is probably an $O_2(\pi^*)$ → $Co(d_{z^2})$ transition in $CoMbO_2$ and $CoHbO_2$.

It would be very useful to be able to monitor the internal modes of bound imidazole, in view of the ubiquitous occurrence of imidazole ligation in hemeproteins, but such modes have not been reported, despite the extensive RR spectroscopy of hemeproteins and imidazole complexes. Both ImH → Fe and Fe → ImH CT transitions should occur somewhere in the u.v.-visible region, but it is difficult to predict their locations, and the intensities may be quite low. A model system is offered by the complex $(ImH)Fe^{III}(CN)_5^{3-}$ [144]. This species has a broad absorption near 480 nm, attributable to ImH → Fe^{III} CT transitions, but the molar absorptivity is only ≃ 300. Excitation in this band does produce resonance enhancement of bound ImH RR modes [145], but only by a small factor. Deprotonation of the bound ImH induces a strong absorption band at 440 nm and a weak band at ≃ 650 nm [144], which may be due to σ and π CT transitions from the bound Im^- [145]. Excitation in these also enhances bound Im^- RR modes, but again the enhancement factors are small [145]. Similar CT transitions are expected for low-spin Fe^{III} hemes with ImH or Im^- ligands, but unless they are appreciably stronger than in the pentacyano complexes, the prospects for monitoring bound imidazole Raman modes are meager.

D. STRUCTURE CORRELATIONS

The first RR spectra of Hb [9, 10] showed differences for different chemical states and gave promise that RR spectroscopy would prove useful in monitoring heme structure. Attention soon focused on bands which appeared to be markers of the oxidation and spin state of the iron [43, 136]. The oxidation-state sensitivity was attributed [43, 60, 61] largely to the backbonding between low-spin Fe^{II} and the porphyrin ring. Removing one of the d_π electrons greatly reduces the backbonding, and low-spin Fe^{III} is probably a net π-acceptor [146].

A continuous change in the oxidation-state marker frequencies can be produced by attaching ligands of graded π-acceptor strength to low-spin Fe^{II} porphyrins [61]. The changes produced by very strong π-acceptors, such as NO and O_2, are as large as those produced by oxidizing Fe^{II} to Fe^{III}.

The spin-state marker bands were originally thought to be responding to a doming of the porphyrin ring, attendant upon the out-of-plane displacement of the iron atom in high-spin heme [43]. Spaulding et al. [147] discovered, however, that one of the spin marker bands, the anomalously polarized band in the 1580-cm^{-1} region (ν_{19}), responded to the porphyrin core size for a range of structurally characterized metalloporphyrins. They suggested core size as the dominant effect in the spin-state sensitivity. This view was subsequently confirmed by further crystallographic [148, 149] and Raman [150–152] work on iron-porphyrin complexes. A negative linear correlation with core size was found for ν_{19} and also for two other spin marker bands, ν_3 and ν_{10} [150–151]. Small but systematic deviations were found for domed or ruffled porphyrins, and an empirical model was proposed to account for these deviations in terms of the loss in porphyrin π-conjugation produced by tilting of the pyrrole rings [150].

i. Core-Size Correlations

When the high-frequency RR modes of protoporphyrin were completely dissected [103, 104], it was found that *all* the skeletal modes above 1450 cm^{-1} show a negative linear dependence on C_t—N, the center-to-pyrrole-N distance of the porphyrin cavity [103]. (The $\simeq$ 1620-cm^{-1} vinyl ν C=C does not show this dependence.) The correlations are shown for a series of protoheme complexes in Figure 15. The distances were obtained from x-ray crystal structures of complexes with similar ligation (see Table 8) but not necessarily the same peripheral substituents on the porphyrin ring. Hoard [153] has noted that the structural parameters for similar complexes are essentially invariant to alterations in the peripheral substituents.

Iron porphyrins [154] can be low- or high-spin, depending on the field of the axial ligands. Strong-field ligands (imidazole, CN^-, CO, O_2) force the valence electrons (Fe^{III} has five, Fe^{II} has six) to pair up in the three d_π orbitals, which point away from the ligands. The bonds to Fe are short, and are only slightly influenced by the oxidation state. When weak-field ligands are bound, the average ligand field is insufficient to overcome the electron pairing energy, the antibonding d_π orbitals are partially occupied, and the Fe—ligand bonds are long. If two equivalent weak-field axial ligands are maintained, the Fe remains in the porphyrin plane, and the porphyrin core expands to accommodate the $d_{x^2-y^2}$ antibonding interaction. The effect is greater for Fe^{II} [149] than for Fe^{III} [148], because the former has lower effective nuclear charge and more extended d-orbitals. The nonbonded interactions between the axial ligands and the pyrrole N atoms are also important in the energetics of core expansion [155].

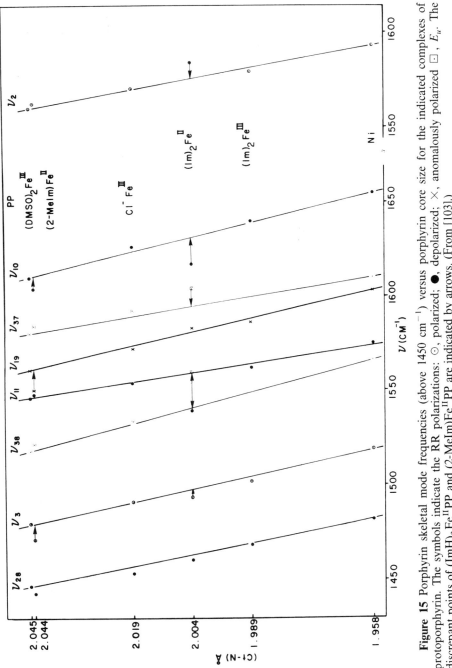

Figure 15 Porphyrin skeletal mode frequencies (above 1450 cm^{-1}) versus porphyrin core size for the indicated complexes of protoporphyrin. The symbols indicate the RR polarizations: ⊙, polarized; ●, depolarized; ×, anomalously polarized ▫, E_u. The discrepant points of (ImH)$_2$Fe^{II}PP and (2-MeIm)Fe^{II}PP are indicated by arrows. (From [103].)

Table 8. Structural Parameters for Iron Porphyrins in Various Ligation, Spin, and Oxidation States

Molecule	Spin state [a]	Coordination number	Distance [b]			Ref.
			$Ct-N_p$	$Fe-N_p$	$Fe-L$	
$Fe^{II}TPP$	is	4	1.972	1.972		158
$(OClO_3^-)Fe^{III}TPP$	is	5	1.979	1.997	2.025	157[e]
$[(ImH)_2Fe^{III}TPP]^+ Cl^-$	ls	6	1.989	1.989	$\left\{\begin{matrix}1.957\\1.991\end{matrix}\right\}$	54
$[C(CN)_3^-]_2Fe^{III}TPP$	is	6	1.995	1.995	2.317	236
$(1\text{-MeIm})_2Fe^{II}TPP$	ls	6	1.997	1.997	2.014	125
$(Cl^-)Fe^{III}TPP$	hs	5	2.012	2.049	2.192	235
$(2\text{-MeImH})Fe^{II}TPP$	hs	5	2.044	2.086	2.161	72
$[(TMSO)_2Fe^{III}TPP]ClO_4^-$ [c]	hs	6	2.045	2.045	$\left\{\begin{matrix}2.069\\2.087\end{matrix}\right\}$	148
$(THF)_2Fe^{II}TPP$ [d]	hs	6	2.057	2.057	2.351	149

[a] ls = low-spin ($S = \frac{1}{2}$ or 0 for Fe^{III} or Fe^{II}), is = intermediate-spin (s = $\frac{3}{2}$ or $\frac{2}{2}$ for Fe^{III} or Fe^{II}), hs = high-spin (s = $\frac{5}{2}$ or $\frac{4}{2}$ for Fe^{III} or Fe^{II})
[b] $Ct-N_p$, center to pyrrole N; $Fe-N_p$, Fe to pyrrole N; $Fe-L$, Fe to axial ligand atom.
[c] TMSO, tetramethylene sulfoxide.
[d] THF, tetrahydrofuran.
[e] Kastner et al.

Since the bonds are weak, the geometry of high-spin hemes is flexible. One of the axial ligands is easily lost, and in the resulting five-coordinated complex, the Fe atoms moves out of the heme plane, by about 0.5 Å [153], allowing the porphyrin core to contract somewhat. [The Fe—pyrrole bonds are shorter in the in-plane than in the out-of-plane structures (see Table 8), reflecting the energy cost of porphyrin expansion.] These bonding patterns determine the locations of the complexes in the correlations of Figure 15. The core-size order is (6-c, ls Fe^{III}) < (6-c, ls Fe^{II}) < (5-c, hs Fe^{III}) < (6-c, hs Fe^{III}, 5-c, hs Fe^{II}) < (6-c, hs Fe^{II}) (c = coordinated; ls, hs = low-, high-spin). Omitted from the graph is the 6-c, hs Fe^{II} example, for which only limited data are available [156]. The two frequencies which have been determined [156], however ($\nu_{10} = 1605$ cm^{-1}, $\nu_{19} = 1551$ cm^{-1}) are within 1 cm^{-1} of the values predicted for them from the correlations.

If the axial ligand field is very weak, e.g. in Fe^{III} perchlorates [157] or for 4-c Fe^{II}, with no axial ligands [158], the spin state is intermediate. The d_{z^2} orbital energy is lowered sufficiently that it participates with the d_π orbitals in sharing the valence electrons, but the $d_{x^2-y^2}$ orbital remains high in energy, and empty. Consequently, the Fe—pyrrole bonds are about as short as they are in low-spin hemes, and the RR frequencies are similar for intermediate- and low-spin complexes of Fe^{II} [61] and Fe^{III} [150, 159].

It is possible to form 5-c ls complexes with very strong-field ligands such as CO and NO. In the case of $(ImH)(NO)Fe^{II}TTP$ [130 (1976)], the loss of the trans axial ImH to give $(NO)Fe^{III}TPP$ [130 (1975)] is accompanied by an out-of-plane displacement (0.2 Å) of the Fe atom, and a slight contraction of the porphyrin core (C_t—N 2.008 → 1.990 Å). These alterations are attributable to nonbonded interactions of the axial ligands with the pyrrole N atoms. RR frequencies are higher for 5-c than for 6-c NO-hemes [62], and the magnitudes of the shifts are consistent with the core-size correlations of Figure 15 [59].

The slopes and intercepts of the correlations are given in Table 9, as are the percentage contributions of methine bond stretching to the potential-energy distribution of the modes, as calculated by Abe et al. [90]. Porphyrin core expansion is expected [13, 147, 150] to be accommodated by weakening of the methine bridge bonds, and the slopes of the correlations are roughly in proportion to the methine involvement in the modes. (The two E_u modes appear to be reversed in this respect; the problem may lie in the normal-mode calculations, since the observed and predicted methine deuteration shifts are also reversed [90].) The slopes and intercepts of the lines in Figure 15 differ somewhat from those obtained previously [150] for ν_{10}, ν_{19}, and ν_3 on the basis of data for a wider range of metalloporphyrins. It seems likely that the correlations are not completely independent of the nature of the metal ion, and they are known [103] to depend somewhat on the peripheral substituents (see below).

Consequently the deviations from the lines which were attributed to doming or ruffing [150] were probably overestimated. For example ClFe^{III}PP, which shows a very slight doming [160] and [(ImH)$_2$Fe^{III}PP]$^+$, whose TPP

Table 9. Iron-Porphyrin Core-Size Correlation Parameters for High-Frequency Skeletal Modes

Mode [a]	P.E.D. [b]	PP [104]			TPP [162]	
		K [c]	A [c]	Δ(ImH)$_2$Fe^{II}PP [d]	K	A
$\nu_{10}(B_{1g})$	C_aC_m (49)	517.2	5.16	−17		
$\nu_{38}(E_u)$	C_bC_b (53) [e]	551.7	4.80	+19		
$\nu_{19}(A_{2g})$	C_aC_m (67)	494.3	5.20	+ 3	526 [f]	4.92
$\nu_3(A_{1g})$	C_aC_m (41)	448.3	5.35	− 5	410 [f]	5.54
$\nu_{28}(B_{2g})$	C_aC_m (52)	402.3	5.64	− 3		
$\nu_2(A_{1g})$	C_bC_b (60)	390.8	6.03	+ 9	262 [f]	7.95
$\nu_{37}(E_u)$	C_aC_m (36) [e]	356.3	6.48	+10		
$\nu_{11}(B_{1g})$	C_bC_b (57)	344.8	6.53	−21		

[a] Mode numbering and symmetries for NiOEP (Abe et al. [90]).
[b] Percentage contribution of the major contributor (C_a—C_m or C_b—C_b stretching) to the NiOEP potential-energy distribution, according to Abe et al. [90].
[c] Slope (cm^{-1}/Å) and intercept (Å) for the relation $\bar{\nu} = K(A - d)$; $\bar{\nu}$ = mode frequency and d = porphyrin center-to-pyrrole nitrogen distance, C_t—N.
[d] Deviations (cm^{-1}) for (ImH)$_2$Fe^{II}PP from the frequencies expected on the basis of its core size.
[e] The calculated ν_{37} and ν_{38} mode compositions are believed to be in error; see Reference [103].
[f] TPP-PP mode correspondences as in Table 6.

analog shows some ruffling [54], both fall on the lines of Figure 15. The only significantly domed structure [72] is probably that of (2-MeIm)Fe^{II}PP, for which the points do fall below the lines, by up to 10 cm^{-1} (see Table 9).

Similar correlations hold for porphyrins other than PP [103], but there are frequency shifts for modes that are coupled to the vinyl groups in PP. Thus both ν_2 and ν_{11}, which are coupled to the vinyl C=C stretch, shift up by ~ 10 cm^{-1} when the vinyl groups are saturated [103], and also, curiously, in heme a [110], where the vinyl and formyl groups are located on opposite sides of the porphyrin ring.

While less work has been done for TPP complexes, dependences of frequency on oxidation and spin state have also been noted for this class [121, 161, 162]. The frequencies for (Fe^{III}TPP)$_2$N are in accord with a low-spin state [163], consistent with XPS results [164] and the structural parmeters [165] in contrast to the high-spin state of [Fe^{III}TPP]$_2$O. Three bands were found to correlate with porphyrin core size [162]. These correspond (see Table 6) to the PP modes ν_2, ν_{19}, and ν_3 (in Reference [162] they were labeled ν_3, ν_4, and ν_6, corresponding to the numbering of bands in Reference [121]), and the slopes of the correlations are in reasonable agreement (Table 9). However, a breakdown of the ν_{19}-like correlation is seen in the data of Chottard et al. [161], which show only a 3-cm^{-1} range for a series of Fe^{II} and Fe^{III} low-spin complexes having a 0.03-Å range of C_t—N distances.

ii. π-Backbonding

The effects of d_π-porphyrin π^*-backbonding are clearly revealed in Figure 15 by the large deviations of the points for (ImH)$_2$Fe^{II}PP. Among those complexes included in the graph, this ls Fe^{II} complex is the only one for which backbonding is expected to be significant. Oxidation to Fe^{III} lowers the d_π-energy, while conversion to the high-spin (2-MeIm)Fe^{II}PP decreases the π^*-d_π overlap due to the lengthened Fe—pyrrole bonds, and the out-of-plane displacement of the Fe.

The deviations for (ImH)$_2$Fe^{II}PP are listed in Table 9. It had been thought [43, 60, 61] that backbonding was associated only with frequency lowerings, which were attributed to the weakening of the porphyrin π-bonds. The deviations are now seen [103], however, to be both positive and negative, as might be expected from the bond alternations that are often associated with π electron shifts. Indeed the deviations, which are symmetry-specific, can be understood qualitatively on the basis of the nodal pattern [53] of the $e_g(\pi^*)$ orbitals, a diagram of which is given in Figure 16. It can be seen that the atomic-orbital phasing is antibonding for one pair of opposite pyrrole C_b—C_b bonds, but bonding for the other pair. Population of this orbital should favor simultaneous stretching of one pair of opposite pyrrole bonds and contraction of the opposite pair. This motion has B_{1g} symmetry, and specific decreases can

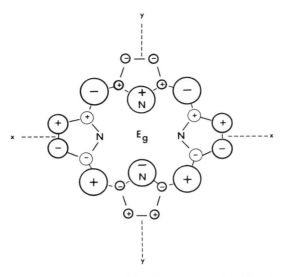

Figure 16 Porphine e_g^* orbital (one of the degenerate pair). The signs refer to the p_z atomic orbitals, and the sizes of the circles are proportional to their coefficients.

be expected for the force constants and frequencies of B_{1g} ring modes. This is indeed observed in the large negative deviations of ν_{10} and ν_{11}.

The E_u modes (ν_{37}, ν_{38}), on the other hand, show large positive deviations. In these modes, stretching of adjacent bonds is accompanied by contraction of opposite ones; this phasing cancels the effects of the e_g^* orbital occupancy and may account for the upshifts. A_{1g} modes (ν_2, ν_3) show smaller deviations, which are both positive and negative. In these modes, all bonds of a given type are expanded or contracted together, a phasing that neither reinforces nor suppresses the e_g modal pattern. Finally, no significant deviations are seen for the $A_{2g}(\nu_{19})$ and $B_{2g}(\nu_{28})$ lines. Because these modes are antisymmetric with respect to the mirror planes which bisect the pyrrole rings, they have no contribution from C_b-C_b stretching.

The π-backbonding interpretation of the low-spin Fe^{II} porphyrin frequencies is supported by the RR spectra obtained by Ksenofontova et al. on the mono- and dianions of zinc and vanadyl etioporphyrin [166], in which the e_g^*-orbitals are partially occupied. The observed frequencies above 1350 cm^{-1} are given in Table 10, and probable assignments are given, based on the band frequencies and polarizations (the E_u modes, ν_{37} and ν_{38}, are not excited, since the activating vinyl groups are absent for EP). It can be seen that adding a pair of electrons to the e_g^*-orbitals produces downshifts of ν_{10}, ν_{11}, and ν_3 comparable to their deviations from the core-size correlations observed for $(ImH)_2Fe^{II}PP$. The expected small upshift in ν_3 is not observed, but the

Table 10. High-Frequency RR Modes (cm^{-1}) of Zinc and Vanadyl Etioporphyrin (EP) and Their Mono- and Dianions [166]

Tentative assignment	VOEP	VOEP$^-$	VOEP^{2-}	$\Delta_{0/2-}$ [a]	ZnEP	ZnEP$^-$	ZnEP^{2-}	$\Delta_{0/2-}$ [a]	$\Delta_{Fe^{II}}$ [b]
ν_{10}	1631		1603	-28	1613	1581	1581	-32	-17
ν_2	1586	1582	1573	-13	1589	1566	1555	-34	$+9$
ν_{11}	1564		1546	-18	1562	1558	1537	-25	-21
ν_3	1497	1484	1475	-22	1484	1484	1475	-9	-5
ν_4	1376	1370	1362	-14	1374	1367	1353	-21	-14 [c]

[a] Change in frequency between the neutral complex and the dianion.
[b] Deviations of $(ImH)_2Fe^{II}PP$ from the core size correlations (Figure 18 and Table 9).
[c] Frequency difference between $(ImH)_2Fe^{II}PP$ and $[(ImH)_2Fe^{III}PP]^+$.

assignment is uncertain in view of the peculiar changes in polarization that occur upon reduction of the EP complexes [166]. The rather irregular changes observed on adding first one, and then two, electrons may be due to symmetry lowering in the (orbitally degenerate) monoanions [166].

iii. Oxidation-State Marker

The $\simeq 1370$-cm^{-1} RR band, which was subsequently identified [29] as the C—N breathing mode ν_4, was recognized in early studies [136, 43] to be an indicator of the oxidation state of the iron atom in hemeproteins. It is found near 1375 cm^{-1} for Fe^{III}-hemes, and near 1360 cm^{-1} for Fe^{II}-hemes, with only small differences between low- and high-spin forms. The increase from Fe^{II} to Fe^{III} was attributed to loss of π-backbonding to porphyrin, in view of the observation that the binding of CO and O_2, both strong π-acceptor ligands, also shifted the frequency up into the Fe^{III} region [43, 60]. Indeed, it subsequently proved possible to adjust ν_4 continuously over the range of Fe^{II}-Fe^{III} frequencies with a graded series of π-acceptor ligands [61]. In this regard, ν_4 behaves like the other high-frequency A_{1g}, B_{1g}, and E_u modes, which are sensitive to the extent of π-backbonding to the porphyrin, as discussed in the preceding section. Moreover, ν_4 also decreases on forming porphyrin mono- and dianions [166], as shown in Table 10.

A particularly interesting instance of ν_4-lowering occurs in the monooxygenase enzyme, cytochrome P450, for which a variety of spectroscopic evidence, including the identification of the Fe—S stretching mode in the RR spectrum [126], implicates cysteine thiolate as the proximal ligand. In the reduced (high-spin Fe^{II}) form, ν_4 is found at $\simeq 1345$ cm^{-1} in proteins from a variety of organisms [167, 168, 31], about 12 cm^{-1} below its frequency in high-spin Fe^{II} hemes with an imidazole ligand. It has been suggested [167, 168] that the lowering reflects strong π-donation from the thiolate ligand. Since the Fe^{II} is high-spin, π-transmission via the d-orbitals seems unlikely, but RS^- has extended p-orbitals, which might interact directly with the porphyrin π^*-orbitals. Such interaction has been invoked [169] to account for the hyperchromic absorption spectrum of the CO adduct; an apparent splitting in the B-band (the characteristic 450-nm component giving cytochrome P450 its name) is attributed to mixing between close lying porphyrin π-π^* and R5 → porphyrin π^* CT transitions. Interestingly, the ground-state interaction is weaker in the CO adduct; ν_4 is at 1368 cm^{-1} [31], only 5 cm^{-1} lower than in HbCO [170] or in a pyridine, CO-heme adduct [61]. Apparently the CO competes effectively with the porphyrin for the RS^- π-electrons, as is also seen in the lowered C—O stretching frequency of cytP450(CO) [171] and of thiolate, CO-heme adducts [172].

In the oxidized form of cytP450, the Fe^{III} is generally low-spin in the absence of substrate, but high-spin when substrate is bound. In the high-spin form, ν_4 is at $\simeq$ 1368 cm^{-1} [167, 168], $\simeq$ 5 cm^{-1} lower than in (Cl$^-$)Fe^{III}PP, which is a good model (see Section E.ii). The low-spin form, however, has a normal value of ν_4, $\simeq$ 1372 cm^{-1}. This is unexpected, since π-donation from RS^- should be particularly effective for a low-spin Fe^{III} complex. If e_g^*-orbital occupancy is the explanation for ν_4 lowering in cytP450, then other bands should also show characteristic shifts, as described in the preceding section. The spectra have yet to be assigned, but it is notable that the reduced protein shows a strong depolarized band at 1534 cm^{-1} [31], which is probably the π^*-sensitive mode $\nu_{11}(B_{1g})$; its frequency is 13 cm^{-1} lower than in high-spin Fe^{II} heme with imidazole as ligand.

The enzyme chloroperoxidase has RR spectra very similar to cytP450 [173]; thiolate ligation is implicated in this case as well.

The e_g^*-orbital occupancy cannot, however, explain all of the observed variations in ν_4. For example, the high-spin Fe^{II} complex (2-MeImH)Fe^{II}PP, a deoxyHb model, shows ν_4 2 cm^{-1} lower than the low-spin (ImH)$_2$Fe^{II}PP (1357 versus 1359 cm^{-1}), although no π-backbonding is expected in the high-spin complex, nor is it observed in the other high-frequency modes (Figure 15). The low ν_4 frequency in (2-MeImH)Fe^{II}PP is not attributable to its expanded core, since the high-spin (DMSO)$_2$Fe^{III}PP, with essentially the same core size, shows ν_4 at a much higher frequency, 1370 cm^{-1}, nearly as high (1373 cm^{-1}) as the low-spin [(ImH)$_2$Fe^{III}PP]$^+$, which has a much smaller core size. An even higher frequency, 1382 cm^{-1}, is seen for the low-spin Fe^{IV} complex found in horseradish peroxidase compound II [174, 175]; this further increase is unlikely to be due to additional reduction in π-backbonding, since Fe^{III} is probably no longer a π-donor, but rather a π-acceptor [146].

It seems likely that the underlying determinant of the oxidation-state sensitivity of ν_4 is polarization of the pyrrole N atoms by the effective nuclear charge, which, in the absence of backbonding, should fall in the order Fe^{IV} > Fe^{III} > Fe^{II}. In the specific case of (ImH)$_2$Fe^{II}PP, there is an increase in the effective nuclear charge due to backbonding, but this is compensated by the porphyrin π^*-orbital occupancy, leaving ν_4 in the Fe^{II} region. When ImH is replaced by π-acceptor ligands, the porphyrin π^*-occupancy is decreased, and ν_4 moves up toward the Fe^{III} region.

E. PROTEIN EFFECTS

As the preceding sections illustrate, many of the basic data on metalloporphyrin RR spectra have been obtained with hemeproteins. The present section focuses on influences on heme RR spectra associated with the protein *per se*. Most of these are attributable to alterations in the axial-ligand bonding among various states of the proteins under study. In principle any direct influence of the polypeptide chain on the porphyrin structure (e.g. skeletal deformations or peripheral substituent conformation changes) should also be reflected in the RR spectrum.

One instance of a direct influence has recently been discovered. The RR spectra of deoxyMb and Mb^{III} (see Figure 13) show bands at 1117 and 1124 cm^{-1}, respectively, which shift *up* upon vinyl C_α deuteration and are therefore assignable to C_b—vinyl stretching (the upshift being due to interactions with the in plane $C_\alpha H$ and $C_\alpha D$ bending modes, which occur at higher and lower frequencies, respectively) [103]. These bands correlate with an i.r.-active E_u mode, ν_{44}, of OEP complexes, and are seen in i.r. but not in RR spectra of protein-free PP complexes. Activation of this mode in the Mb RR spectra indicates a strong symmetry-lowering effect. It has been suggested [103] to be due to an electrostatic field, generated by charged groups and dipoles, localized nare the heme vinyl groups, similar to the one calculated for Hb by Warshel and Weiss [58].

i. Metal-Ligand Modes

a. Fe—NO, Fe—O_2. The nitrosyl group has a strong labilizing effect on the *trans* ligand, as seen in NO-heme structural parameters [130], and the proximal imidazole bond of hemeprotein nitrosyl adducts is subject to weakening and rupture. Alone among low-spin adducts of human Hb A, HbNO can be switched from the R to the T quaternary state by the addition of the effector molecule inositol hexaphosphate (IHP) [176]. This switch is accompanied by spectral changes that have been interpreted as reflecting breaking of the Fe—ImH bonds in half of the subunits [176]. Consistent with this inference, an additional Fe—NO stretching frequency is observed at a higher frequency, as expected for five-coordinated NO-heme (Stong et al. [62]). The same frequency shift is seen for MbNO when the pH is lowered from 8.4 to 5.8, indicating that the Fe—ImH bond is broken at low pH, presumably via protonation of the proximal imidazole [59]. In the case of carp HbNO, however, no shift in the Fe—NO band is seen upon IHP addition, despite the known change in quaternary structure, indicating that no Fe—ImH bonds are broken in this protein; likewise there are no changes in the porphyrin skeletal modes [186].

HbO$_2$ is not switched to the T-state by IHP addition, but this switch apparently can be accomplished in the mutant hemoglobins Hb Kansas and Hb Milwaukee [177, 178]. Nagai et al. [128] have examined the O$_2$ adducts of these mutant Hb's, and report no changes greater than 3 cm^{-1} in the Fe—O$_2$ frequency upon IHP addition. Apparently the Fe—O$_2$ force constant, and by implication the Fe—O$_2$ bond strength, is essentially undiminished in the T-state, despite the lowered O$_2$ affinity. The Fe—O$_2$ frequency has also been investigated [122, 123] for the "picket fence" O$_2$-heme analogs (O$_2$)(B)Fe^{II}T$_{piv}$PP [179], where B = 1-MeIm, 2-MeImH, or 1,2-diMeIm, in the solid state and in solution, and at various temperatures. The sterically hindered imidazoles lower the O$_2$ affinity substantially, providing a model for T-state Hb [179], and the crystal structure of the 2-MeImH adduct [180] shows the Fe—O$_2$ bond to be significantly stretched, by 0.15 Å relative to the 1-MeIm adduct [181]. The Fe—O$_2$ frequency also decreases, but only by 10 cm^{-1} at low temperature in the solid state [122], and the frequency difference diminishes in solution. On the basis of Badger's rule [182] only a ~ 0.01-Å increase would be expected for a frequency lowering of 10 cm^{-1}, but Badger's rule might be inapplicable in sterically hindered molecules [122]. The connection, if any, between the O$_2$ affinity and the Fe—O$_2$ bond strength remains cloudy. (Affinity changes can, of course, be due to free-energy differences in the deoxy form.)

Tsubaki et al. [68] found that IHP addition had no effect on the Fe—CO frequency of HbCO Kansas, but did produce an asymmetric broadening of the band of carp HbCO, suggesting partial conversion to a conformer with lower ν Fe—CO.

b. Fe—ImH. The Fe—ImH mode of deoxyHb, at $\simeq 220$ cm^{-1}, has recently attracted much attention because it is altered appreciably between R and T states of Hb's which are chemically modified to permit this switch [128]. In an elegant experiment, Nagai and Kitagawa [183] showed, with the aid of reconstituted valency hybrids of Hb, that the broad and asymmetric Fe—ImH band in deoxyHb A contains separate contributions from the β (220 cm^{-1}) and α (207 cm^{-1}) chains. Both of these shift up, to 224 and 222 cm^{-1}, respectively, when the T quaternary constraints are relaxed by chemical modification.

On the assumption that these frequency shifts result from mechanical tension on the Fe—ImH bonds, Nagai and Kitagawa calculated the associated bond extensions and strain energies, assumed to be the energy required to displace the atoms from their equilibrium positions along a Morse potential; their estimates of strain energies were only 31 and 4 cal/mol for β and α chains, much less than thermal energies at room temperature. If, however, the shifts are interpreted as reflecting changes in the bond strength, then substantial energy changes can be estimated, 1.3 and 0.4 kcal/mol for α and β chains [127], which are an appreciable fraction of the free energy of cooperativity [184] $\simeq$ 3.4 kcal/mol. On the basis of comparable frequency shifts between aqueous and nonaqueous solutions of the 2-MeImH model compound, it has been suggested [127] that the lowered T-state frequencies might be associated with decreased Fe—ImH bond strengths, due to decreased N_ε H-bonding to the backbone carbonyl group. Protein crystal structures [142] are not inconsistent with this hypothesis, although the observed differences are of marginal significance in relation to the uncertainty in the coordinates. Altered N_ε H-bonding has previously been suggested as a mechanism for controlling the O_2 affinity [185]; increased H-bonding should increase electron donation from ImH to Fe, which in turn should increase the affinity for the π-acceptor ligand O_2.

It has been pointed out [127] that weakening the Fe—ImH bond should raise the free energy of deoxyHb, and therefore raise, rather than lower, the O_2 affinity, if all other factors are constant. Lowering the O_2 affinity depends on amplifying the Fe—ImH bond weakening in HbO_2; this is expected to occur in either the model based on mechanical tension (which should increase as the Fe atom is forced into the heme plane) or that based on the H-bond free energy (which should increase as the O_2 ligand draws electron density from the heme). A critical test would be to monitor the Fe—ImH stretching frequency in R- and T-state HbO_2. Unfortunately, the prognosis for this experiment is not good, because of the weakness of this mode in the RR spectrum; as noted in Section C.iii.a, it has tentatively been identified for MbO_2 as a weak shoulder at $\simeq 270$ cm^{-1} [59].

Interestingly, IHP addition to deoxyHb from carp shifts the Fe—ImH band from 223 to 214 cm^{-1} [186]. The band remains symmetric in the T-state, and its frequency is halfway between those of the α and β chains in T-state deoxyHb A. Evidently the Fe—ImH bonds in carp deoxyHb are weakened to an intermediate degree in both α and β chains. This is consistent with the absence of Fe—ImH bond breaking in T-state carp HbNO [186].

c. Fe^{III}—F, Fe^{III}—OH. The bands assigned to Fe—F and Fe—OH stretching have been found by Asher and Schuster [79] to show different frequencies for Hb^{III} and Mb^{III} (495 versus 490 cm^{-1} for Fe—OH; 471, 443 versus 461, 422 cm^{-1} for Fe—F). These frequency lowerings were interpreted as reflecting weakening of the Fe—OH and Fe—F bonds in Mb^{III}, due to increased displacement of the Fe atom toward the proximal imidazole. Variability in the Fe—F frequencies was also observed [187] for isolated α (466, 441 cm^{-1}) and β (471, 444 cm^{-1}) chains. However, there is no discernable shift in the Fe—F frequencies upon IHP addition to $Hb^{III}F$, which is expected to switch the quaternary structure from R to T [78, 187]. This observation is consistent with the apparent lack of sensitivity of the Fe—O_2 frequency to the quaternary state [128]. Interestingly, large downshifts of the Fe—F frequency are observed [188] at lower pH for both $Hb^{III}F$ (461 → 399 cm^{-1}) and $Mb^{III}F$ (468 → 407 cm^{-1}), with pK_a's of 5.1 and 5.5 respectively. These shifts were interpreted in terms of protonation of the distal imidazole, with H-bonding of the resulting imidazolium ion to the bound fluoride [188].

ii. Porphyrin Skeletal Modes

Changes in axial ligation can also alter the porphyrin skeletal frequencies through steric and/or electronic effects, as discussed in Section D. Most of the protein-induced shifts in the high-frequency heme RR bands can be interpreted in this way.

Thus the large RR frequency differences between aquo-metHb and native HRP, both of which are high-spin ferriheme-containing proteins, was originally attributed to differences in porphyrin doming [189]; but the subsequently established correlations between the skeletal frequencies and porphyrin size and ligation showed that the differences were those expected for six- versus five-coordinated high-spin Fe^{III} heme [150]. Other spectroscopic results also support the inference that native HRP does not have a water molecule directly bound to the iron atom [190, 191]. Exogenous ligands do nevertheless bind, and the fluoride complex [189], although still high-spin, has the same, lowered skeletal frequencies as are observed for fluoro- or aquo-metHb, consistent with six-coordination.

The intermediate-pH (9-11) form of oxidized cytochrome c', called form II, which also contains high-spin ferriheme, likewise shows frequencies characteristic of five-coordination [150, 192-194]. The recently reported crystal structure of cytochrome c' [195] confirms the absence of a sixth ligand. However, LaMar et al. [196] attribute a pair of proton NMR resonances of form II to the methylene group of a sixth ligand. At lower pH (5-9), another form, I, is stable. Its ESR spectrum has been interpreted [197] in terms of a quantum mixture of intermediate high-spin states. The RR spectra [150, 192-194] are consistent with this interpretation, since the skeletal frequencies move toward values characteristic of intermediate-spin Fe^{III}. Since this state is observed only for very weak axial ligands, such as percholorate, it appears that the axial ligand bond is weakened at lower pH. This is consistent with the inference from NMR linewidths that the ligand field is weaker in I than in II, although loss of a sixth ligand would also have this effect. Recent Mössbauer [198] and near-I.R. MCD [199] data, however, favor a high-spin state for I.

The substrate-bound form of cytochrome P450 has skeletal frequencies characteristic of 5-coordination, as shown by a comparison with a model complex, (p-nitrolenzenethiolate)$Fe^{III}PP$ [200]. The high-frequency shifts observed in the HbNO RR spectrum upon $R \to T$ conversion [62, 176, 201, 202] result from the dissociation of the proximal imidazoles in half the chains, as confirmed by comparison with model complexes. Nagai et al. [62] used the meso-deuterium sensitivity of the skeletal frequencies to establish, via selective reconstitution with labeled heme, that Fe—ImH bond rupture occurs in the α-chains, as had previously been conjectured [176]. The six → five-coordination shifts have recently been examined in detail for MbNO [59] and found to be consistent with the contraction in porphyrin core size, 0.018 Å, observed [130] between (NO)(1-MeIm)Fe^{II}TPP and (NO)Fe^{II}TPP, reflecting the relief of nonbonded interactions upon release of the sixth ligand; some bands did not shift as expected, implying a superimposed electronic effect associated with the stronger interaction (shorter Fe—NO bond [130]) in the five-coordinated complex [59].

In the case of Fe^{III} cytochrome c, Kitagawa et al. [203] observed frequency increases of 3, 4, and 5 cm^{-1}, in bands that can now be identified as ν_{10}, ν_{19}, and ν_{11}, upon raising the pH from 7.8 to 12.4. It is well established that this pH change results in the replacement of met80 with another ligand, probably a lysine, and the RR shifts presumably reflect this ligand substitution. The observed shifts suggest an increase in the porphyrin core size by $\simeq 0.01$ Å, which is plausibly related to lessened nonbonded interaction with the porphyrin N atoms due to the lower steric requirement of $-NH_2$ than $-SCH_3$ as a ligand.

Asher and Schuster [79] made the interesting observation that the low-spin components of Hb^{III}(OH$^-$) and Mb^{III}(OH$^-$) give rise to RR bands, which can be identified with ν_{10} and ν_{19}, which in turn show substantial frequency differences, 8 and 6 cm^{-1}, in the direction indicating a smaller porphyrin core size for Mb^{III}(OH$^-$) (by $\simeq 0.014$ Å, according to the C_t—N correlations). The high-spin components, however, showed essentially the same frequencies for ν_{10} and ν_{11}.

Experiments designed to examine the effect of Hb quaternary structure on the porphyrin skeletal modes have shown no significant differences between R and T states [204, 205] except for HbNO (see above) in which the Fe—ImH bond is broken in the T-state, or for Hb^{III}(H$_2$O) [206, 207], in which the T-state has a higher fraction of high-spin heme. Using a Raman difference technique [20], however, Rousseau and coworkers [207–210] have detected systematic R-T frequency differences of 0.3–2 cm^{-1} in certain skeletal modes. In the deoxy form, Hb that is modified chemically to destabilize the T structure shows shifts to lower frequencies in ν_4, and also ν_3, ν_{11}, ν_2, and ν_{10} [209]. High-spin Hb^{III} adducts, however, show downshifts in ν_4 when the T-state is stabilized by IHP addition [207]. Noting the dependence of ν_4 on the porphyrin π^*-orbital occupancy, Rousseau and coworkers suggested a charge-transfer origin of the R-T shift in ν_4, with nearby phenylalanine side chains transferring charge to the heme, and more so in the R than in the T state [209]. They suggested further that the charge-transfer process may modulate O$_2$ binding, via the polarization of the heme by the bound O$_2$, and thereby provide the link between quaternary structure and O$_2$ affinity [209]. This interpretation, however, is open to the objection that the filled π-orbitals of phenylalanine are much lower in energy than the porphyrin π^*-orbitals and significant charge transfer is unlikely. There are alternative mechanisms that could produce small frequency shifts, e.g. electrostatic polarization of the porphyrin [58] by the protein, or protein-induced geometry changes. Perutz [211] has suggested that the ν_4 frequency lowering in the R-state of deoxyHb may be associated with the strengthening of the Fe—ImH bonds, as evidenced by the raised Fe—ImH frequencies (see Section E.i.b). The expected increase in nonbonded interactions associated with the stronger Fe—ImH bonds might explain the observed lowerings [209] of ν_3, ν_{11}, ν_2, and ν_{10}, all of which decrease with increasing porphyrin core size. Small differences in skeletal mode frequencies have also been observed for cytochrome c from different organisms [210].

F. TIME-RESOLVED STUDIES

The application of time-resolved resonance Raman spectroscopy holds great promise in studies of hemeprotein dynamics, because of the high structural information content of RR spectra. The time evolution of the RR spectrum can give more specific insights into the structural changes in reacting systems than can the more conventional absorption or fluorescence spectra. The advent of pulsed lasers and optical multichannel detectors has made time-resolved RR spectroscopy possible on time scales down to picoseconds. As with other forms of kinetic spectroscopy, a variety of impulses can be used to initiate the desired reaction, but laser photolysis naturally commends itself, since a laser is already in use to obtain the RR spectrum. In the simplest experiment, those which have been most widely reported to date, the same laser pulse is used to photolyze the sample and produce the RR spectrum. In this case the RR spectrum corresponds to the average composition of the sample over the duration of the pulse. Increasingly, pulse-probe arrangements are being devised, in which one set of pulses is used to photolyze the sample and the second set to generate the RR spectrum, providing a time slice of the evolving system at the delay interval between the pulses. Care must be taken that the probe pulses do not themselves produce significant photolysis. In either case the spectrum is usually built up over many repeated laser pulses, and the interval between the photolysis pulses must be sufficiently long to allow chemical regeneration of the system, assuming it to be reversible, or else the sample must be renewed between pulses, usually by flowing. It is also possible to obtain submicrosecond or longer time resolution with cw lasers, simply by flowing the sample through the beam.

The photolability of HbCO makes it a natural object of photolysis studies. Three time-resolved Raman investigations of this molecule on the 5–10-ns time-scale appeared nearly simultaneously [212–214]. In one, a N_2-dye laser system was used to generate the CARS spectrum of the photoproduct [212], while in the other two, a frequency-doubled YAG laser was used to produce the spontaneous Raman spectrum [213, 214]. In all three studies the photoproduct spectrum was reported to be quite similar to the deoxyHb spectrum, but Lyons et al. [214] reported small frequency shifts for bands known to be sensitive to the porphyrin core size (see Section D.i). Subsequent investigation with a synchronously pumped mode-locked dye laser produced a photoproduct spectrum within 30 ps [215, 216], which showed the same small frequency shifts relative to deoxyHb (Figure 17). It was known that photoexcitation of HbCO gives a very fast absorption change [217–219] (< 11 ps), presumably associated with CO dissociation, and that the resulting absorption spectrum, which resembles that of deoxyHb but is broader [219], persists beyond 680 ps.

The RR frequency shifts were interpreted as being due to the iron atom remaining closer to the heme plane than in deoxyHb, resulting in a slightly expanded porphyrin core [215, 216]. The extent of Fe-atom displacement could not be accurately gauged, because of the smallness of the frequency shifts, but the same frequencies were observed [216] for a high-spin six-coordinated

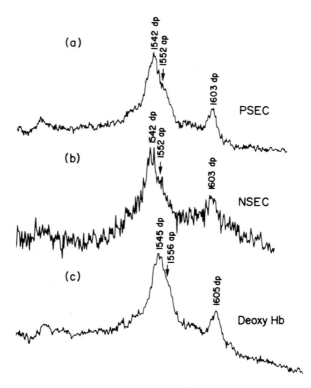

Figure 17 RR spectra, obtained with 5760-Å pulsed excitation from a synchronously pumped dye laser, of deoxyHb (c) and of the HbCO photoproduct obtained with ≃ 30-ps (a) and ≃ 20-ns (b) pulses. See Reference [216] for details. (From reference 215, with frequencies redetermined in reference 216.)

Fe^{II}-heme, $(THF)_2Fe^{II}PP$ (THF = tetrahydrofuran), whose TPP analog is known to have the Fe atom lying in the porphyrin plane [149]. The frequency shifts, relative to deoxyHb, persisted beyond 20 ns [215, 216], but were shown in a subsequent cw-flow experiment to relax within 300 ns [220]. This relaxation was attributed [216] to a tertiary conformation change of the protein, involving movement of the F-helix, to which the Fe atom is attached via the hisF80 proximal imidazole, which preceeds the $R \to T$ quaternary-structure change, occurring in the $1-100$-μs time scale [221, 222].

Since the difference between low-spin HbCO and high-spin deoxyHb is sensitively reflected in the core-size marker frequencies, the prompt appearance of a deoxy-like photoproduct spectrum implied that the spin-state change occurred within 30 ps of photoexcitation, and must have occurred via intersystem crossing to an excited ligand-field state of HbCO [216], since thermal spin conversion of Fe^{II} complexes has been found to occur on a much longer time scale, 10–100 ns [223]. Although this mechanism had been suggested previously [217], the RR spectrum provided the first structural evidence for a prompt spin conversion.

Similar results were subsequently obtained for HbO_2; with $\simeq 50$-ps synchronously pumped dye-laser pulses (5 nJ) the photoproduct spectrum was the same, within experimental error, as that obtained for HbCO [224]. Excitation of HbO_2 with shorter ($\simeq 30$ ps) and stronger (0.2 mJ) pulses from a frequency-doubled mode-locked YAG laser produced a different spectrum, however [225] (Figure 18). The frequencies of bands identifiable with ν_{10} and ν_{11} were 10 and 5 cm^{-1} lower than in the COHb photoproduct. This altered spectrum could be identified with the prompt absorption spectral transient seen by Chernoff et al. [226] for HbO_2, but not for HbCO, which decayed within 90 ps. The large RR frequency shifts were tentatively attributed [225] to the formation of a deoxyHb electronically excited state with π-π* character, in view of the known sensitivity of ν_{10} and ν_{11} to π*-orbital occupancy (see Section E.ii). Cornelius et al. [227] had suggested the intermediacy of a deoxyMb electronically excited state for MbO_2 photolysis, to explain the observation of a common absorption spectral transient upon photolysis of either MbO_2 or deoxyMb, with intensity greater for MbO_2. Also, 30-ps YAG-excited RR spectra were reported by Coppey et al. [228]. Although some photolysis was observed, with a deoxyHb-like contribution, the spectra had insufficient resolution or signal/noise to allow determination of the photoproduct spectrum.

The $\simeq 300$-ns cw-flow experiment [220] also showed a Fe—ImH stretching band characteristic of R-state deoxyHb. The same shift to a symmetric band at 223 cm^{-1} from an asymmetric band with components at 220 cm^{-1} (β-chains) and 207 cm^{-1} (α-chains) was seen in the flow experiment (Figure 19), as upon chemical modification, to release the T-state constraints [128]. Similar results have been obtained with 10-ns YAG laser photolysis [229]. Thus the kinetic experiments fully support the R-T differences in the Fe—ImH stretching frequency which had been deduced by chemical means [128, 183].

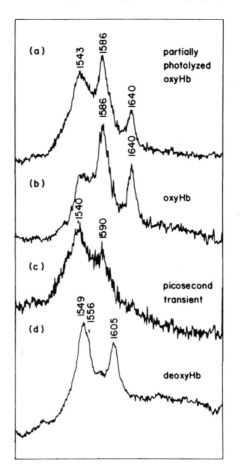

Figure 18 RR spectra obtained with 5320-Å, $\simeq$ 30-ps pulses from a mode-locked YAG laser, of deoxyHb (d) and of the HbO_2 photoproduct (c) obtained by subtracting the spectrum (b) of HbO_2 (loose laser focus) from that (a) of a photolysis mixture (tight laser forces). See Reference [225] for details. (From [225].)

Photolysis of HbCO also produces a prompt (10-ns) frequency for ν_4 which is 2 cm^{-1} lower than in deoxyHb [230]; this shift persists beyond 300 ns [220]. Pulse-probe experiments by Lyons and Friedman [231] indicate that the relaxation of this shift is biphasic. The smaller-amplitude ($\simeq$ 0.5 cm^{-1}) second phase with a $\simeq$ 60-μs relaxation time corresponds to the $R \rightarrow T$ conversion, while the larger amplitude ($\simeq$ 1.5 cm^{-1}) first phase has a 0.8-μs relaxation time, and must be due to a tertiary change. This change, however, is different from that associated with the core-size markers; as noted above, the latter relax within 0.3 μs [220]. The nature of this second tertiary relaxation is unknown.

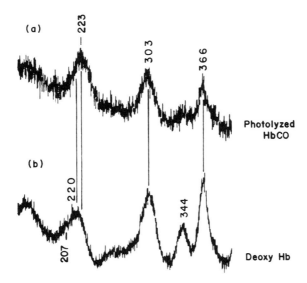

Figure 19 Low-frequency RR spectra, obtained with cw 4545-Å excitation, of deoxyHb and of the HbCO photoproduct generated during the $\simeq 0.3$-μs transit time of the sample stream in the laser. The R-T alteration of the Fe—ImH stretching mode, at $\simeq 220$ cm^{-1}, is revealed in this kinetic experiment. See Reference [220] for details. (From [220].)

Presumably it is not connected with Fe—ImH bond alteration, since the latter seems to be associated with the quaternary structure. (However, the Fe—ImH band has not yet been monitored in the vicinity of the 0.8-μs relaxation time.)

Pulse radiolysis has been used to reduce cytochrome c via hydrated electrons for time-resolved RR studies [232]. At neutral pH, the RR spectrum of cyt^{II} c was observed within 1 μs, and there was no subsequent change. At alkaline pH, however, the initial frequency of ν_{11} is 1533 cm^{-1}, and it relaxes to its normal cyt^{II} c value, 1547 cm^{-1}, with a time constant longer than 20 ms but shorter than a few seconds. Since the same lowered frequency of ν_{11} had been observed [233] for carboxymethylated cyt^{II} c, in which met80 is prevented by alkylation from binding to the Fe atom, and since at high pH cyt^{III} c is known to have met80 replaced by another ligand, probably a lysine, the authors inferred that they were monitoring the religation of met80 upon reduction of high-pH cyt^{III} c. The slowness of the process presumably reflects the required protein conformation change.

Acknowledgments

Drs. S. A. Asher, M. A. El-Sayed, J. M. Friedman, R. M. Hochstrasser, D. L. Rousseau, W. M. Woodruff, and N.-T. Yu kindly communicated preprints of their work prior to publication. Dr. S. Schoi's help in preparing the manuscript is greatly appreciated. Work from the author's laboratory described herein was supported by Grant HL 12526 from the National Institute of Health.

References

1. Raman, C. V., and Krishnan, K. S. *Nature, Lond.* **121** (1928) 501.
2. Krishnan, R. S. *Indian J. Pure and Appl. Phys.*, "Raman Effect Golden Jubilee Number", **16** (1978) 129.
3. Hibben, J. H. *The Raman Effect and Its Chemical Application*, Reinhold, New York, 1939.
4. Kiefer, W. In *Advances in Infrared and Raman Spectroscopy* Vol. 3, R. J. H. Clark and R. E. Hester (eds.), Heyden, London, 1977, p. 1.
5. Spiro, T. G., and Stein, P. *Ann Rev. Phys. Chem.* **28** (1977) 501.
6. Spiro, T. G., and Gaber, B. P. *Ann Rev. Biochem.* **46** (1977) 553.
7. Carey, P. R., and Salares, B. R. In *Advances in Infrared and Raman Spectroscopy*, Vol. 7, R. J. H. Clark and R. E. Hester (eds.), Heyden, London, 1980, p. 1.
8. Spiro, T. G., and Loehr, T. M. In *Advances in Infrared and Raman Spectroscopy*, Vol. 1, R. J. H. Clark and R. E. Hester (eds.), Heyden, London (1975), Chapter 3.
9. Strekas, T. C., and Spiro, T. G. *Biochim Biophys. Acta* **263** (1972) 830.
10. Brunner, H., Mayer, A., and Sussner, H. *J. Mol. Biol.*, **70** (1972) 153.
11. Spiro, T. G., and Strekas, T. C. *Proc. Natl. Acad. Sci.* **69** (1972) 2622.
12. Spiro, T. G., *Biochim. Biophys. Acta* **416** (1975) 169. Spiro, T. G. *Proc. Roy. Soc. Lond. A* **345** (1975) 89.
13. Warshel, A. *Annual Revs. Biophys. and Bioeng.* **6** (1977) 273.
14. Felton, R. H., and Yu, N.-T. In *The Porphyrins*, Vol. 3, Part A, D. Dolphin (ed.), Academic, New York, 1978, pp. 347–388.
15. Kitagawa, T., Ozaki, Y., and Kyogoku, Y. *Advances in Biophysics* **11** (1978) 153.
16. Spiro, T. G. In *Methods in Enzymology, Biomembranes, Part E*, S. Fleischer and L. E. Packer (eds.), Academic, 1978, p. 233. Spiro, T. G. *Israel J. Chem.* **21** (1981) 81.
17. Asher, S. A. In *Methods in Enzymology*, A. Eraldo, L. B. Bernardi, E. Cheancone (eds.), Academic, 1981.
18. Tang, J., and Albrecht, A. C. In *Raman Spectroscopy*, H. A. Szimanski (ed.), Plenum, New York, 1970, Vol. 2, p. 33.
19. Johnson, B. B., and Peticolas, W. L. *Ann. Rev. Phys. Chem.* **27** (1976) 465.
20. Hennoker, W. H., Penner, A. P., Siebrand, W., and Zgierski, M. *J. Chem. Phys.* **69** (1978) 1704.
21. Mortensen, O. S. *Chem. Phys. Lett.* **30** (1975) 406.
22. Small, G. J., and Yeung, E. S. *Chem. Phys.* **9** (1975) 375.
23. Tonks, D. L., and Page, J. B. *Chem. Phys. Lett.* **79** (1981) 247.
24. Makinen, M. W., and Churg, A. K. In *Physical BioInorganic Chemistry* A. B. P. Lever and H. B. Gray (eds.), Addison-Wesley, Reading, Mass., 1982,
25. Simpson, W. T. *J. Chem. Phys.* **16** (1948) 1124; **17** (1949) 1218.
26. Gouterman, M. in *The Porphyrins*, D. Dolphin (ed.), Academic, New York, Vol. III, Part A, pp. 1–156. 1979
27. Verma, A. L., and Bernstein, H. J. *J. Raman Spectrosc.* **2** (1974) 163. Verma, A. L., and Bernstein, H. J. *J. Chem. Phys.* **61** (1974) 2560. Mendelsohn, R., Sunder, S., Verma, A. L., and Bernstein, H. J. *J. Chem. Phys.* **62** (1975) 37. Verma, A. L., Mendelsohn, R., and Bernstein, H. J. *J. Chem. Phys.* **61** (1974) 383. Verma, A. L., Asselin, M., Sunder, S., and Bernstein, H. J. *J. Raman Spectrosc.* **4** (1975) 295.
28. Strekas, T. C., Packer A., and Spiro, T. G. *J. Raman Spectrosc.* **1** (1973) 197.

29. Kitagawa, T., Abe, M., Kyogoku, K., Ogoshi, H., Sugimoto, H., and Yoshida, Z. *Chem. Phys. Lett.* **48** (1977) 55.
30. Dutta, P. K., Dallinger, R., and Spiro, T. G. *J. Chem. Phys.* **73** (1980) 3580.
31. Champion, P. M., Gunsalus, J. C., and Wagner, G. C. *J. Am. Chem. Soc.* **100** (1978) 3473. Remba, R. D., Champion, P. M., Fitchen, D. B., Chiang, R., and Hager, L. P. *Biochemistry* **18** (1979) 2280.
32. Champion, P. M., and Albrecht, A. C. *J. Chem. Phys.* **71** (1979) 1110.
33. Warshel, A., and Dauber, P. *J. Chem. Phys.* **66** (1977) 5477. Liang, R., Schnepp, O., and Warshel, A. *Chem. Phys.* **34** (1978) 17.
34. Shelnutt, J. A. Cheung, L. D. Chang, R. C. C. Yu, N-T. and Felton, R. H. *J. Chem. Phys.* **66** (1977) 3387.
35. Burke, J. M., Kincaid, J. R., and Spiro, T. G. *J. Am. Chem. Soc.* **100** (1978) 6077.
36. Champion, P. M., and Albrecht, A. C. *J. Chem. Phys.* **72** (1980) 6498.
37. Shelnutt, J. A. *J. Chem. Phys.* **72** (1980) 3948. Champion, P. M., and Albrecht, A. C. *J. Chem. Phys.* **75** (1981) 3211.
38. Champion, P. M., and Lange, R. *J. Chem. Phys.* **73** (1980) 5947.
39. Perrin, M. H., Gouterman, M., and Perrin, C. L. *J. Chem. Phys.* **50** (1969) 4137.
40. Kitagawa, T., Ogoshi, H., Watanabe, E., and Yoshida, Z. *J. Phys. Chem.* **79** (1975) 2629.
41. McClain, W. M. *J. Chem. Phys.* **55** (1971) 2789.
42. Pezolet, M., Nafie, L. A., and Peticolas, W. J. *J. Raman Spectrosc.* **1** (1973) 455. Nestor, J. R., and Spiro, T. G. *J. Raman Spectrosc.* **1** (1973) 539.
43. Spiro, T. G., and Strekas, T. C. *J. Am. Chem. Soc.* **96** (1974) 338.
44. Warshel, A. *Chem. Phys. Lett.* **43** (1976) 273.
45. Sunder, S., Mendelsohn, R., and Bernstein, H. J. *J. Chem. Phys.* **63** (1975) 573.
46. Strekas, T. C., and Spiro, T. G. *J. Raman Spectrosc.* **1** (1973) 387.
47. Collins, D. W., Champion, P. M., and Fitchen, D. B. *Chem. Phys. Lett.* **40** (1976) 416.
48. Shelnutt, J. A., O'Shea, D. C., Yu, N.-T., Cheung, L. D., and Felton, R. H. *J. Chem. Phys.* **64** (1976) 1156.
49. Friedman, J. M., and Hochstrasser, R. M. *Chem. Phys.* **1** (1973) 457. Mingardi, M., and Siebrand, W. *J. Chem. Phys.* **62** (1975) 1074.
50. Shelnutt, J. A., O'Shea, D. C., Yu, N.-T., Cheung, D., and Felton, R. H. *J. Chem. Phys.* **64** (1976) 1156.
51. Choi, S., Spiro, T. G., (submitted for publication).
52. Ogoshi, H., Masai, N., Yoshida, Z., Takemoto, T., and Nakamoto, K. *Bull. Chem. Soc. Japan* **44** (1971) 49.
53. Longuett-Higgins, H. C., Rector, C. W., and Platt, J. R. *J. Chem. Phys.* **18** (1950) 1174.
54. Collins, D. M., Countryman, R., and Hoard, J. L. *J. Am. Chem. Soc.* **94** (1972) 2066.
55. Asher, S., and Sauer, K. *J. Chem. Phys.* **64** (1976) 4115.
56. Yu, N.-T., and Tsubaki, M. *Biochemistry*, **19** (1980) 4647.
57. Takano, T. *J. Mol. Biol.* **110** (1977) 537, 569.
58. Warshel, A., and Weiss, R. M. *J. Am. Chem. Soc.* **103** (1981) 446.
59. Walters, M. A., and Spiro, T. G. *Biochemistry* (in press).
60. Kitagawa, T., Iizuka, T., Saito, M., and Kyogoku, Y. *Chem. Lett.* (1975) 849.

61. Spiro, T. G., and Burke, J. M. *J. Am. Chem. Soc.* **98** (1976) 5482.
62. Stong, J. D., Burke, J. M., Daly, P., Wright, P., and Spiro, T. G. *J. Am. Chem. Soc.* **102** (1980) 5815. Nagai, K., Welborn, C., Dolphin, D., and Kitagawa, T. *Biochemistry* **19** (1980) 4755.
63. Brunner, H. *Naturwiss.* **61** (1974) 129.
64. Makinen, M. W., and Eaton, W. A. *Ann. N.Y. Acad. Sci.* **206** (1973) 210.
65. Eaton, W. A., Hanson, L. K., Stephens, P. J., Sutherland, J. C., and Dunn, J. B. R. *J. Am. Chem. Soc.* **100** (1978) 4991.
66. Churg, A. K., and Makinen, M. W. *J. Chem. Phys.* **68** (1978) 1913; **69** (1978) 2668. Makinen, M. W., Churg, A. K., and Glick, H. A. *Proc. Natl. Acad. Sci.* **75** (1978) 2291.
67. Chottard, G., and Mansuy, D. *Biochem. Biophys. Res. Commun.* **77** (1977) 1333.
68. Armstrong, R. S., Irwin, M. J., and Wright, P. *J. Amer. Chem. Soc.* **104** (1982) 626. Tsubaki, M., Srivastava, R. B., and Yu, N.-T. *Biochemistry*, **21** (1982) 1132.
69. Mitchell, M., Choi, S., and Spiro, T. G. Manuscript in preparation. Desbois, A. and Lutz, M. *Biochim. Biophys. Acta* **671** (1981) 257.
70. Kincaid, J., Stein, P., and Spiro, T. G. *Proc. Natl. Acad. Sci.* **76** (1979) 549, 4156.
71. Kitagawa, T., Nagai, K., and Tsubaki, M. *FEBS Lett.* **104** (1979) 376.
72. Hoard, J. L., and Scheidt, W. R. *Proc. Natl. Acad. Sci.* **70** (1973) 3919; **71** (1974) 1578.
73. Kincaid, J., and Nakamoto, K. *Spectrosc. Lett.* **9** (1976) 19.
74. Kitagawa, T., Abe, M., Kyogoku, Y., Ogoshi, H., Watanabe, G., and Yoshida, Z. *J. Phys. Chem.* **80** (1976) 1181.
75. Boucher, L. J. *Coord. Chem. Rev.* **7** (1972) 289.
76. Gaughan, R. R., Shriver, D. F., and Boucher, L. J. B. *Proc. Natl. Acad. Sci.* **72** (1975) 433.
77. Shelnutt, J. A., O'Shea, D. C., Yu, N.-T., Cheung, L. D., and Felton, R. H. *J. Chem. Phys.* **64** (1976) 1156.
78. Asher, S. A., Vickery, L. E., Shuster, T. M., and Sauer, K. *Biochemistry* **16** (1977) 5849.
79. Asher, S. A., and Schuster, T. M. *Biochemistry* **18** (1979) 5377.
80. Desbois, A., Lutz, M., and Banerjee, R. *C. R. Hebd. Seances Acad. Sci., Ser. D* **287** (1978) 349. Desbois, A., Lutz, M., and Banerjee, R., *Biochemistry* **18** (1979) 1510.
81. Tsubaki, M., Srivastava, R. B., and Yu, N.-T. *Biochemistry* **20** (1981) 946.
82. Eaton, W. A., and Hochstrasser, R. M. *J. Chem. Phys.* **49** (1968) 985.
83. Lanir, A., Yu, N.-T., and Felton, R. H. *Biochemistry* **18** (1979) 1656.
84. Wright, P. G., Stein, P., Burke, J. M., and Spiro, T. G. *J. Am. Chem. Soc.* **101** (1979) 3531.
85. Kobayashi, H., and Yanagawa, Y. *Bull. Chem. Soc. Japan* **45** (1972) 450.
86. Gaber, B. P., Miskowski, V., and Spiro, T. G. *J. Am. Chem. Soc.* **96** (1974) 6868.
87. Tsubaki, M., and Yu, N.-T. *Proc. Natl. Acad. Sci.* **78** (1981) 3581–3585.
88. Nakamoto, K., Suzuki, M., Ishiguro, T., Kozuka, M., Nishida, Y., and Kida, S. *Inorg. Chem.* **19** (1980) 2822.
89. Kitagawa, T., Abe, M., and Ogoshi, H. *J. Chem. Phys.* **69** (1978) 4516.
90. Abe, M., Kitagawa, T., and Kyogoku, Y. *J. Chem. Phys.* **69** (1978) 4526.
91. Ogoshi, H., Saito, Y., and Nakamoto, K. *J. Chem. Phys.* **57** (1972) 4194.
92. Stein, P., Burke, J. M., and Spiro, T. G. *J. Am. Chem. Soc.* **97** (1975) 2304.

93. Abe, M., Kitagawa, T., and Kyogoku, Y. *Chem. Lett.* (1976) 249.
94. Sunder, S., and Bernstein, H. *J. Raman Spectrosc.* **5** (1976) 351.
95. Susi, H., and Ard, J. S. *Spectrochim. Acta* **A33** (1977) 561.
96. Warshel, A., and Lappicirella, A. *J. Am. Chem. Soc.* **103** (1981) 4664–4673.
97. Hoard, J. L. In *Porphyrins and Metalloporphyrins*, K. N. Smith (ed.), Elsevier, New York, 1975, pp. 317–376.
98. Larrabee, J. A., and Spiro, T. G. *J. Am. Chem. Soc.* **102** (1980) 4217.
99. Bürger, H., Burezyk, K., and Furhop, J. H. *Tetrahedron* **27** (1971) 3257.
100. Desbois, A., Momenteau, M., Loock, B., and Lutz, M. *Spectrosc. Lett.* **14** (1981) 257–269.
101. Ogoshi, H., Watanabe, E., Yoshida, Z., Kincaid, J., and Nakamoto, K. *J. Am. Chem. Soc.* **95** (1973) 2845.
102. Colthup, N. B., Daly, L. H., and Weiberley, F. E. *Infrared and Raman Spectroscopy*, 2nd ed., Academic, New York, 1975, p. 275.
103. Choi, S., Spiro, T. G., Langry, K. C., Smith, K. N., Budd, D. L., and La Mar G. N. *J. Am. Chem. Soc.* (in press).
104. Choi, S., Spiro, T. G., Langry, K. C., and Smith, K. N. *J. Am. Chem. Soc.* (in press).
105. Malmstrom, B. G. In *Metal Ion Activation of Dioxygen*, T. G. Spiro (ed.), Wiley, New York, 1980, pp. 181–208.
106. Salmeen, I., Rimai, L., Gill, D., Yamamoto, T., Palmer, G., Hartzell, C. R., and Beinert, H. *Biochem. Biophys. Res. Commun.* **52** (1973) 1100–1107.
107. Kitagawa, T., Kyogoku, Y., and Orii, Y. *Arch. Biochem. Biophys.* **181** (1977) 228–235.
108. Salmeen, I., Rimai, L., and Babcock, G. *Biochemistry* **17** (1978) 800.
109. Adar, F., and Yonetani, T. *Biochim. Biophys. Acta.* **502** (1978) 80–86.
110. Ondrias, M. R., and Babcock, G. T. *Biochem. Biophys. Res. Commun.* **93** (1980) 29–35.
111. Bocian, D. F., Lemley, A. T., Petersen, N. O., Brudvig, G. W., and Chan, S. I. *Biochemistry* **18** (1979) 4396.
112. Woodruff, W. H., Dallinger, R. F., Antalis, T. M., and Palmer, G. *Biochemistry* **20** (1981) 1332–1338.
113. Callahan, P. M., and Babcock, G. T. *Biochemistry* **20** (1981) 952. Babcock, G. T., Callahan, P. M., Ondrias, M. R., and Salmeen, I. *Biochemistry* **20** (1981) 952–966.
114. Woodruff, W. H., Kessler, R. H., Ferris, N. S., Dallinger, R. F., Carter, K. R., Antalis, T. M., and Palmer, G. *Adv. Chem.* in press.
115. Steelandt-Frentrup, J. V., Salmeen, I., and Babcock, G. G. *J. Am. Chem. Soc.* **103** (1981) 591–592.
116. Choi, S., Lee, J. J., Wei, Y. H., and Spiro, T. G. Submitted for publication.
117. Zwarich, R., Smolarek, J. and Goodman, L. *J. Mol. Spectrosc.* **38** (1971) 336–357. Green, J. H. S., and Harrison, D. J. *Spectrochim. Acta* **32a** (1976) 1265–1277.
118. Fuchsman, W. H., Smith, Q. R., and Stein, M. M. *J. Am. Chem. Soc.* **99** (1977) 4190.
119. Eaton, S. S., and Eaton, G. R. *J. Am. Chem. Soc.* **97** (1975) 3660.
120. La Mar, G. M., Eaton, G. R., Holm, R. H. and Walker, F. A. *J. Am. Chem. Soc.* **95** (1973) 63.
121. Burke, J. M., Kincaid, J. R., Peters, S., Gagne, R. R., Collman, J. P., and Spiro, T. G. *J. Am. Chem. Soc.* **100** (1978) 6083.

122. Walters, M. A., Spiro, T. G., Suslick, K. S., and Collman, J. P. *J. Am. Chem. Soc.* **102** (1980) 6857.
123. Hori, H., and Kitagawa, T. *J. Am. Chem. Soc.* **102** (1980) 3608–3613.
124. Tsubaki, M., Nagai, K., and Kitagawa, T. *Biochemistry* **19** (1980) 379–385.
125. Mashiko, T., Reed, C. A., Haller, K. J., Kastner, M. E., and Scheidt, W. R. *J. Am. Chem. Soc.* **103** (1981) 5758–5767.
126. Champion, P. M., Stallard, B. R., Wagner, G. C., and Gunsalus, I. C. *J. Am. Chem. Soc.*, in press.
127. Stein, P., Mitchell, M., and Spiro, T. G. *J. Am. Chem. Soc.* **102** (1980) 7795.
128. Nagai, K., Kitagawa, T., and Morimoto, H., *J. Mol. Biol.* **136** (1980) 271–289.
129. Teraoka, J., and Kitagawa, T. *Biochem. Biophys. Res. Commun.* **93** (1980) 694–700. Teraoka, J., and Kitagawa, T. *J. Biol. Chem.* **256** (1981) 3969–3977.
130. Scheidt, W. R., and Piciulo, P. L. *J. Am. Chem. Soc.* **98** (1976) 1913. Scheidt, W. R., and Frisse, M. E. *J. Am. Chem. Soc.* **97** (1975) 17.
131. Weiss, J. *Nature* **203** (1964) 83.
132. Pauling, L. *Nature*, **203** (1964) 182.
133. Wayland, B. B., Minkiewicz, J. V., and Abd-Elmageed, M. E. *J. Am. Chem. Soc.* **96** (1974) 2795. Wayland, B. B., and Abd-Elmageed, M. E. *J. Am. Chem. Soc.* **96** (1974) 9809.
134. Reed, C. A., and Chung, S. K. *Proc. Natl. Acad. Sci.* **74** (1977) 1780–1784.
135. Drago, R. S., and Corden, B. B. *Accts. Chem. Res.* **13** (1980) 353–360.
136. Yamanoto, T., Palmer, G., Gill, D., Salmeen, I. T., and Rimai, L. *J. Biol. Chem.* **248** (1973) 5211.
137. Barlow, C. H., Maxwell, J. C., Wallace, W. J., and Caughey, W. S. *Biochem. Biophys. Res. Commun.* **55** (1973) 91. Maxwell, J. C., Volpe, J. A., Barlow, C. H., and Caughey, W. S. *Biochem. Biophys. Res. Commun.* **58** (1974) 166. Collman, J. P., Brauman, J. I., Halbert, B. R., and Suslick, K. S. *Proc. Natl. Acad. Sci.* **73** (1976) 3333.
138. Case, D. A., Huynh, B. H., and Karplus, M., *J. Am. Chem. Soc.* **101** (1979) 4433.
139. Nakamoto, K. *Infrared and Raman Spectra of Inorganic and Coordination Compounds*, 3rd ed., Wiley-Interscience, New York, 1978, pp. 212–213.
140. Collman, J. P., and Reed, C. A. *J. Am. Chem. Soc.* **95** (1973) 2048–2049.
141. Anet, F. A. L., Basus, V. J., Heutt, A. P. W., and Saunders, M. *J. Am. Chem. Soc.* **102** (1980) 3945.
142. Bolton, W., and Perutz, M. F. *Nature* **228** (1970) 551–552. Ladner, R. C., Heidner, E. J., and Perutz, M. F. *J. Mol. Biol.* **114** (1977) 385–414.
143. Mincey, T., and Traylor, T. G. *J. Am. Chem. Soc.* **101** (1979) 765–766.
144. Shepherd, R. E. *J. Am. Chem. Soc.* **98** (1976) 3329.
145. Walters, M. A., and Spiro, T. G. Submitted for publication.
146. Shulman, R. G., Glarum, S. H., and Karplus, M. *J. Mol. Biol.* **57** (1971) 93.
147. Spaulding, L. D., Chang, C. C., Yu, N.-T., and Felton, R. H. *J. Am. Chem. Soc.* **97** (1975) 2517.
148. Mashiko, T., Kastner, N. E., Spartalian, K., Scheidt, W. R., and Reed, C. A. *J. Am. Chem. Soc.*, **100** (1978) 6354–6362.
149. Reed, C. A., Mashiko, T., Scheidt, W. R., Spartalian, K., and Lang, G. *J. Am. Chem. Soc.* **102** (1980) 2302–2306.
150. Spiro, T. G., Stong, J. D., and Stein, P. *J. Am. Chem. Soc.* **101** (1979) 2648.
151. Huong, P. V., and Pommier, J.-C. *C. R. Acad. Sci., Ser. C* **285** (1977) 519.
152. Scholler, D. M., and Hoffman, B. M. *J. Am. Chem. Soc.* **101** (1979) 1655.

153. Hoard, J. L. In *Porphyrins and Metalloporphyrins*, K. M. Smith (ed.), American Elsevier, New York, 1975, pp. 317-376.
154. Scheidt, W. R., and Reed C. A. *Chem. Rev.* **81** (1981) 543.
155. Olafson, B. D., and Goddard, W. A. *Proc. Natl. Acad. Sci.* **74** (1977) 1315. Warshel, A. *Proc. Natl. Acad. Sci.* **74** (1977) 1789.
156. Terner, J., Stong, J. D., Spiro, T. G., Nagumo, M., Nicol, M., and El-Sayed, M. A. *Proc. Natl. Acad. Sci.* **78** (1981) 1313-1317.
157. Dolphin, D. H., Sams, J. R., and Tsin, T. G. *Inorg. Chem.* **16** (1977) 711. Kastner, M. E., Scheidt, W. R., Mashiko, T., and Reed, C. A. *J. Am. Chem. Soc.* **100** (1978) 666.
158. Collman, J. P., Hoard, J. L., Kim, N., Land, G., and Reed, C. A. *J. Am. Chem. Soc.* **97** (1975) 2676.
159. Teraoka, J., and Kitagawa, T. *J. Phys. Chem.* **84** (1980) 1928-1935.
160. Koenig, D. F. *Acta Cryst.* **18** (1965) 63.
161. Chottard, G., Battioni, P., Battioni, J.-P., Lange, M., and Mansuy, D. *Inorg. Chem.* **20** (1981) 1718-1722.
162. Stong, J. D., Kubaska, R. J., Shupack, S. I., and Spiro, T. G. *J. Raman Spectrosc.* **9** (1980) 312-314.
163. Schick, G. A., and Bocian, D. F. *J. Am. Chem. Soc.* **102** (1980) 7982; *J. Amer. Chem. Soc.* (submitted).
164. Kadish, K. M., Bottomley, L., Brace, J. G., and Winograd, N. J. *J. Am. Chem. Soc.* **102** (1980) 4341.
165. Scheidt, W. R., Summerville, D. A., and Cohen, I. A. *J. Am. Chem. Soc.* **98** (1976) 6623.
166. Ksenofontova, N. M., Maslov, V. G., Sidorov, A. N., and Bobovich, Ya. S. *Opt. Spectrosc.* **40** (1976) 462-465.
167. Champion, P. M., and Gunsalus, I. C. *J. Am. Chem. Soc.* **99** (1977) 2000.
168. Ozaki, Y., Kitagawa, T., Kyogoku, Y., Shimada, H., Iizuka, T., and Ishimoura, Y. *J. Biochem.* **80** (1976) 1447. Ozaki, Y., Kitagawa, T., Kyogoku, Y., Imai, Y., Hashimoto-Yutfudo, C., and Sato, R. *Biochemistry* **17** (1978) 5826.
169. Hanson, L. K., Eaton, W. A., Sligar, F. J., Gunsalus, I. C., Gouterman, M., and Connel, C. R. *J. Am. Chem. Soc.* **98** (1976) 2672.
170. Rimai, L., Salmeen, I. T., and Petering, D. H. *Biochemistry* **14** (1975) 378.
171. Bohm, S., Rein, H., Janig, G.-R., and Ruckpaul, K. *Acta Biol. Med. Germ.* **35** (1976) 27.
172. Dawson, J. H., Holm, R. H., Trudell, J. R., Barth, G., Linder, R. E., Bunnenberg, E., Djerassi, C., and Tang, S. C. *J. Am. Chem. Soc.* **98** (1976) 3707.
173. Champion, P. M., Remba, R. D., Chaing, R., Fitchen, D. B., and Hager, L. P. *Biochim. Biophys. Acta* **446** (1976) 486. Remba, R. D., Champion, P. M., Fitchen, D. B., Chaing, R., and Hager, L. P. *Biochemistry* **18** (1979) 2280.
174. Rakhit, G., Spiro, T. G., and Uyeda, M. *Biochem. Biophys. Res. Commun.* **71** (1976) 803.
175. Felton, R. H., Romans, A. Y., Yu, N.-T. and Schonbaum, G. R. *Biochim. Biophys. Acta* **434** (1976) 82.
176. Perutz, M. F., Kilmartin, J. V., Nagai, K., Szabo, A., and Simon, S. R. *Biochemistry* **15** (1976) 378, and references therein.
177. Ogawa, S., Mayer, A., and Shulman, R. G. *Biochem. Biophys. Res. Commun.* **40** (1972) 1224.
178. Fung, L. W.-M., and Ho, C. *Biochemistry* **14** (1975) 2526.

179. Collman, J. P., Braumann, J. I., Rose, E., and Suslick, K. S., *Proc. Natl. Acad. Sci.* **75** (1978) 1052.
180. Jameson, G. B., Molinaro, F. S., Ibers, J. A., Collman, J. P., Braumann, J. I., Rose, E., and Suslick, K. S. *J. Am. Chem. Soc.* **100** (1978) 6769; **102** (1980) 3224.
181. Collman, J. P., Gagne, R. R., Reed, C. A., Robinson, W. T., and Rodley, G. A. *Proc. Natl. Acad. Sci.* **71** (1974) 1326. Jameson, G. B., Robinson, W. G., Gagne, R. R., Reed, C. A., and Collman, J. P. *Inorg. Chem.* **17** (1978) 850.
182. Hershbach, D. R., and Laurie, V. W. *J. Chem. Phys.* **35** (1961) 458.
183. Nagai, K., and Kitagawa, T. *Proc. Natl. Acad. Sci.* **77** (1980) 2033–2037.
184. Perutz, M. F. *Br. Med. Bull.* **32** (1976) 195–207.
185. Blumberg, W. E., and Peisach, J. *Adv. Chem. Ser.* **100** (1971) 271–291. Valentine, J. S., Sheridan, R. P., Allen, L. C., and Kahn, P. *Proc. Natl. Acad. Sci.* **76** (1979) 1009–1013.
186. Walters, M. A., and Spiro, T. G. To be published.
187. Asher, S. A., and Schuster, T. M. *Biochemistry* **20** (1981) 1866–1873.
188. Asher, S. A., Adams, M. L., and Schuster, T. M. *Biochemistry*, **20** (1981) 3339–3346.
189. Rakshit, G., and Spiro, T. G. *Biochemistry* **13** (1974) 5317–5323.
190. Lanir, A., and Schejter, A. *Biochem. Biophys. Res. Commun.* **62** (1975) 199–203.
191. Kobayashi, K., Tamura, M., Hayshi, K., Hori, H., and Morimoto, H. *J. Biol. Chem.* **255** (1979) 2239–2242.
192. Strekas, T. C., and Spiro, T. G., *Biochim. Biophys. Acta* **351** (1974) 237.
193. Kitagawa, T., Ozaki, Y., Kyogoku, Y., and Horio, T. *Biochim. Biophys. Acta* **493** (1977) 1–11.
194. Teraoka, J., and Kitagawa, T. *J. Phys. Chem.* **84** (1980) 1928–1935.
195. Weber, P. C., Howard, A., Xuong, N. H., and Salemme, F. R. *J. Mol. Biol.* **153** (1981) 399.
196. La Mar, G. N., Jackson, J. T., and Bartsch, R. G. *J. Amer. Chem. Soc.* **103** (1981) 4405–4410.
197. Maltempo, M. M., Moss, T. H., and Cusanovich, M. A. *Biochim. Biophys. Acta* **342** (1974) 290–305.
198. Emptage, M. H., Zimmermann, R., Que, L., Jr., Munck, E., Hamilton, W. D., and Orme-Johnson, W. H. *Biochim. Biophys. Acta* **495** (1977) 12–23.
199. Rawlings, J., Stephens, P. J., Jafie, L. A., and Kamen, M. D. *Biochemistry* **16** (1977) 1725–1729.
200. Anzenbacher, P., Sipal, Z., Strauch, B., Twardowski, J., and Proniwicz, L. M. *J. Amer. Chem. Soc.* **103** (1981) 5929–5930.
210. Szabo, A., and Barron, L. D. *J. Amer. Chem. Soc.* **97** (1975) 17.
202. Scholler, D. M., Hoffman, B. M., and Shriver, D. F. *J. Amer. Chem. Soc.* **98** (1976) 7866.
203. Kitagawa, T., Ozaki, Y., Tereoka, Y., Kyogoku, Y., and Yamanaka, T. *Biochim. Biophys. Acta* **494** (1977) 100.
204. Sussner, H., Mayer, A. and Brunner, J. *Eur. J. Biochem.* **41** (1974) 465.
205. Scholler, D. M., Hoffman, B. M., and Shriver, D. R. *J. Amer. Chem. Soc.* **98** (1976) 7866.
206. Ferrone, F. A., and Topp, W. *Biochem. Biophys. Res. Commun.* **66** (1975) 444.
207. Rousseau, D. L., Shelnutt, J. A., Henry, E. R., and Simon, S. R. *Nature* **285** (1980) 49–51.

208. Rousseau, D. L. *J. Raman Spectrosc.* **10** (1981) 94.
209. Shelnutt, J. A., Rousseau, D. L., Friedman, J. M., and Simon, S. R. *Proc. Natl. Acad. Sci. U.S.A.* **76** (1979) 4409–4413.
210. Shelnutt, J. A., Rousseau, D. L., Dethmers, J. K., and Margoliash, E. *Proc. Natl. Acad. Sci. U.S.A.* **76** (1979) 3865.
211. Perutz, M. F. *Proc. Roy. Soc. London B* **208** (1980) 135.
212. Dallinger, R. F., Nestor, J. R., and Spiro, T. G. *J. Amer. Chem. Soc.* **100** (1978) 6251.
213. Woodruff, W. H., and Farquharaon, S. *Science* **201** (1978) 831.
214. Lyons, K. B., Friedman, J. M., and Fleury, P. A. *Nature* **275** (1978) 565.
215. Terner, J., Spiro, T. G., Nagumo, M., Nicol, M. F., and El-Sayed, M. A. *J. Amer. Chem. Soc.* **102** (1980) 3238.
216. Terner, J., Stong, J. D., Spiro, T. G., Nagumo, M., Nicol, M. F., and El-Sayed, M. A. *Proc. Natl. Acad. Sci. U.S.A.* **78** (1981) 1313–1317.
217. Shank, C. V., Ippen, E. P., and Bersohn, R. *Science* **193** (1976) 50–51.
218. Noe, L. J., Eisert, W. G., and Rentzepis, P. N. *Proc. Natl. Acad. Sci. U.S.A.* **75** (1978) 573–577.
219. Greene, B. I., Hochstrasser, R. M., Weisman, R. W., and Eaton, W. A. *Proc. Natl. Acad. Sci. U.S.A.* **75** (1978) 5255–5259.
220. Stein, P., Terner, J., and Spiro, T. G., *J. Phys. Chem.*, **86** 168 (1982).
221. Sawicki, C. A., and Gibson, Q. H. *J. Biol. Chem.* **251** (1975) 1533.
222. Cho, K. C., and Hopfield, J. J. *Biochemistry* **18** (1979) 5826.
223. Beattie, J. K., Sutin, N., Turner, D. H., and Flynn, T. W. *J. Amer. Chem. Soc.* **95** (1973) 2052–2054. Beattie, J. K., Binstead, R. A., and West, R. J. *J. Amer. Chem. Soc.* **100** (1978) 3044–3050.
224. Nagumo, M., Nicol, M., and El-Sayed, M. A. *J. Phys. Chem.* **85** (1981) 2431–2438.
225. Terner, J., Voss, D. F., Paddock, C., Miles, R. B., and Spiro, T. G. *J. Phys. Chem.*, **86** 859 (1982).
226. Chernoff, D. A., Hochstrasser, R. M., and Steele, A. W. *Proc. Natl. Acad. Sci. U.S.A.* **77** (1980) 5606–5610.
227. Cornelius, P. A., Steele, W. A., Chernoff, D. A., and Hochstrasser, R. M. *Proc. Natl. Acad. Sci. U.S.A.* (1982).
228. Coppey, M., Tourbez, H., Valat, P., and Alpert, B. *Nature* **284** (1980) 568–570.
229. Irwin, M. J., and Atkinson, G. H. *Nature* **293** (1981) 317.
230. Friedman, J. M., and Lyons, K. B. *Nature* **284** (1980) 570.
231. Lyons, K. B., and Friedman, J. M. In *Symposium on Interactions Between Iron and Proteins in Oxygen and Electron Transport*, C. Ho and W. A. Eaton, (eds.), Elsevier, 1981.
232. Cartling, B. E., and Wilbrandt, R. *Biochim. Biophys. Acta* **637** (1975) 61–68.
233. Ikeda-Saito, M., Kitagawa, T., Iizuka, T., and Kyogoku, Y. *FEBS Lett.* **50** (1975) 233–235. Kitagawa, T., Kyogoku, Y., Iizuka, T., Ikeda-Saito, M., and Yamanaka, T. *J. Biol. Chem.* **78** (1975) 719–728.
234. Yu, N.-T., and Srivastava, R. B. *J. Raman Spectrosc.* **9** (1980) 166–171.
235. Hoard, J. L., Cohen, G. H., Glick, M. D. *J. Amer. Chem. Soc.* **89** (1967) 1992.
236. Summerville, D. A., Cohen, I. A., Hatano, K., and Scheidt, W. R. *Inorg. Chem.* **17** (1978) 2906–2910.

4
The Electrochemistry of Iron Porphyrins In Nonaqueous Media

KARL M. KADISH

A. INTRODUCTION

Electrochemical techniques for measuring porphyrin redox potentials have been used for some time. The early work was initially done by potentiometry and is summarized in the reviews by Falk [1] and Clark [2]. In these studies the midpoint potentials for a variety of metalloporphyrins, including those of iron, were reported. These measurements were usually carried out in aqueous media containing a variety of axial ligands. In the mid 1960s several research groups began to use the techniques of polarography and cyclic voltammetry to measure the electrode reactions of iron porphyrins. Most electrochemical techniques used in recent years include cyclic voltammetry and polarography as well as measurements at a rotating solid electrode (usually platinum) or measurements by differential pulse polarography. In theory, if the electrode reactions are rapid (called reversible in electrochemical terminology), there should be no difference between the half-wave potentials obtained using any of these electrochemical techniques. This is often the case, but for complexes where the electron transfer is slow in comparison with mass transport, (called a quasireversible or irreversible electron transfer), $E_{1/2}$ may not approximate E^0 and it is not always a simple matter to know if the measured half-wave potential is close to the thermodynamically significant standard potential. Examples for this will be given in the discussion of each technique.

In this review no electrochemistry of heme proteins will be discussed, and only results on synthetic and naturally occurring porphyrins will be considered. In addition, the presentation will be limited almost exclusively to electrochemical reactivity in nonaqueous media. The electrochemistry of iron porphyrins in aqueous solutions has been discussed in detail in the older literature as well as in several more recent reviews. This review will only consider these reactions briefly and thus might be considered to summarize the electrochemistry of "model compounds" in nonaqueous media.

The models which have been used most frequently were initially those which were readily available (the naturally occurring porphyrins) and those which were the most easily synthesized (tetraphenylporphyrins and octaethyl porphyrins); only recently have complexes been synthesized for investigating specific structure-function relationships (such as picket-fence porphyrins). In addition to synthetic considerations, solubility has also been the problem for a number of complexes. Most electrochemical techniques require solution concentrations of 10^{-3} to 10^{-4} M, and partly for this reason nonaqueous solvents have been utilized.

A list of the specific substituents for naturally occurring porphyrins and for several synthetic porphyrins has been presented on page 4 of volume 1 of this book. The nature of the substituent is important in determining the basicity of the conjugated porphyrin system and will directly influence half-wave potentials and rates of electron transfer for the electrooxidation-reduction of each complex.

The stable oxidation state of synthetic iron porphyrins is iron(III). Iron(II) porphyrins are easily oxidized to the iron(III) form, and for this reason great care must be taken in removing all traces of residual oxygen from solutions in which iron(II) is generated. Autooxidation of iron(II) porphyrins often gives a five-coordinated dimeric complex with a bridging central oxygen atom, μ-oxo bis(porphyrin iron(III))dimer. Depending on the solvent, these dimeric complexes may be quite stable and will exhibit redox properties quite different from those of the monomeric species. Although a number of electrochemical studies of μ-oxo iron(III) dimers have been published, it is still not completely clear where the added or abstracted electron(s) are localized.

The most studied electron-transfer reaction of iron porphyrins has been Fe(III) $\rightleftharpoons$ Fe(II). In fact, prior to 1965 this was the only reaction which had been investigated. The reason for this is that almost all early electrochemical studies of porphyrins were undertaken in aqueous media, where the reactions at extreme cathodic or anodic potentials are dominated by the electrochemistry of the solvent.

Starting in 1965, Stanienda [3, 4] began to use nonaqueous media (in this case butyronitrile) to observe the oxidation of a number of metalloporphyrin complexes to either a higher metal oxidation state or a cation radical. In these studies were included the complex protoFeF which was reversibly oxidized by two one-electron transfers beyond that of the Fe(III) complex. Six years later Felton and coworkers [5, 6] characterized the first step as producing Fe(IV). At about the same time, Felton and Linschitz [7] and Clack and Hush [8] were measuring the reduction potentials of a number of metalloporphyrins in DMSO and butyronitrile and observed that both π anion radicals and dianions could be produced. However, no iron complexes were included in these extensive studies. It was not until the early 1970s that a number of laboratories began to investigate, in detail, the electrode reactions of iron porphyrins in nonaqueous media.

In 1974–1975 Lexa and coworkers [9, 10] investigated the reduction of TPPFeX in a number of nonaqueous solvents containing nitrogeneous bases. The laboratory of Davis looked at similar complexes from 1975–1978, and the laboratory of Kadish investigated both monomeric and dimeric species from 1973 until 1982. Recently, a number of other groups have reported potentials and characterization of the oxidation and reduction products of synthetic iron porphyrin systems. It is now clear that the synthetic iron(III) porphyrin complex may be oxidized or reduced by up to six electrons without destroying the conjugated ring system. Further irreversible reductions may be observed when the porphyrin ring is protonated.

The electroreductions starting with monomeric iron(III) may be represented as occurring in four discrete one-electron-transfer steps, of which two

Introduction

involve transitions at the central metal ion, and two at the ring:

$$[PFe(III)]^+ \overset{e}{\rightleftharpoons} PFe(II) \overset{e}{\rightleftharpoons} [PFe(I)]^- \overset{e}{\rightleftharpoons} [PFe(I)]^{-2}$$
$$\updownarrow e$$
$$[PFe(I)]^{-3}$$

The electrooxidation of iron(III) may be represented as occurring in two one-electron-transfer steps, of which the first may or may not involve the central metal ion. Thus, the sequence can be represented as follows:

$$[PFe(III)]^+ \begin{matrix} \overset{-e}{\nearrow} [PFe(IV)]^{+2} \overset{-e}{\rightleftharpoons} [PFe(IV)]^{+3} \\ \underset{-e}{\searrow} [PFe(III)]^{+2} \overset{-e}{\rightleftharpoons} [PFe(III)]^{+3} \end{matrix}$$

Although the assignment of oxidation states to the central metal is quite clear for complexes of Fe(III) and Fe(II), it is difficult to discern the degree of delocalization in the other steps. This has been discussed extensively for the complexes of Fe(IV) and to some extent for those of Fe(I).

The following pages will deal with the factors influencing both the oxidation and the reduction of Fe(III) and Fe(II) porphyrins. During the last decade several laboratories, including those of the author, have been interested in the electrode reactions of iron porphyrins. The main thrust of these studies has been to correlate changes in half-wave potentials ($E_{1/2}$) and electron-transfer rates (k^0) with changes in the electron-transfer mechanisms and pathways, and at the same time, to elucidate structure-function relationships. With regard to the latter studies, attempts have been made to correlate $E_{1/2}$ and k^0 with the following:

(a) metal spin state,
(b) iron-porphyrin-plane distance,
(c) axial-ligand coordination,
(d) solvent system,
(e) counterion on Fe(III),
(f) basicity of the porphyrin ring.

Because of the difficulty in separating these influences, there is not always a clear-cut answer. However, some insights have been gained by these studies and these will be presented.

B. TECHNIQUES

i. Polarography

Polarography is the name applied to potential sweep voltammetry at a dropping mercury electrode (DME) [11]. Voltammetry comprises a group of electroanalytical procedures that are based upon the measurement of current-potential curves at a polarized electrode in the solution of interest.

Recent years have seen an immense growth in the use of polarographic techniques as well as the development of several new methods closely related to polarography. The application of polarographic analysis to organic and biological systems has been extensive in Europe for some years, but it is only recently that this application has found widespread use in North America. Since this technique is now commonly used and is discussed in detail in most analytical and instrumental-analysis textbooks, a general technical discussion will not be presented here. However, details sufficient to understand the technique as applied to the analysis of porphyrins will be given.

The potential range of a DME is such that reductions but not oxidations of the neutral metalloporphyrin may be observed. The most positive anodic limit of the DME varies with solvent and supporting electrolyte, but under the best conditions is not greater than $+0.5$ V versus SCE. This limit is due to oxidation of the mercury electrode itself. Thus no cation radical formation of metalloporphyrins may be observed on mercury. The negative range of potential for a DME will also depend on the solvent and supporting electrolyte, and will range from approximately -1.0 V in an $HClO_4$ aqueous solution to -2.8 V versus SCE in DMF or DMSO.

To obtain the best current-voltage curves from polarography a solution of approximately 10^{-3} to 5×10^{-4} M is utilized, and to this an inert electrolyte is added. This electrolyte (called the supporting electrolyte) is usually 10^{-1} M to 1 M and is used to prevent migration of the electroactive species toward the electrode [12]. In nonaqueous media, the electrolyte is also necessary to minimize the resistance of the solution, which may be quite high. There is no need to have any particular volume of solution, and the volume is usually dictated by the cost of the porphyrins and the solvents. For hard-to-obtain or expensive complexes, solution volumes of a few microliters have been utilized. Under normal conditions the usual volume is 8–15 ml, which allows the use of conventional-size glassware. Early instrumentation involved a two-electrode system, a reference electrode, usually saturated calomel (SCE), and a DME as the working electrode. All modern instruments now employ a three-electrode configuration, in order to minimize resistance and IR loss in the solution [13].

Figure 1 illustrates a typical polarogram obtained for the reduction of TPPFeCl in CH_2Cl_2, 0.1 M TBAP. Of interest are the height of the wave, i_d, (the current at the plateau); the shape of the wave, which tells something about the rate of electron transfer; and the position of the inflection point ($E_{1/2}$) on

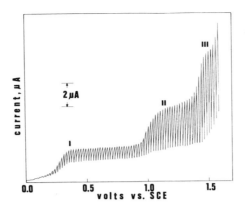

Figure 1 Polarogram obtained for reduction of TPPFeCl in CH_2Cl_2, 0.1 M TBAP.

the wave, which tells something about the standard potential for reduction of the complex if the process is reversible.

For the specific example illustrated in Figure 1 there are three reductions waves, which (we know) correspond to the successive one-electron reduction of first Fe(III), then Fe(II), and finally Fe(I). The heights of the three waves are equal, implying an equal number of electrons transferred in each step. The exact equation for the diffusion limiting current maximum, i_d, is given by the Ilkovic equation and is

$$i_d = 708nCD^{1/2}m^{2/3}t^{1/6}, \tag{1}$$

where i_d is the current in μA, n the number of electrons transferred, C the concentration in mM, D the diffusion coefficient in cm^2/sec, m the flow rate of mercury in mg/sec, and t the drop time in seconds. A natural drop time is usually 3–6 seconds, but with most commercial instrumentation now available, an automatic and constant drop time is possible, usually on the order of 0.3 to 10 seconds. The average current, often reported in the older literature, is usually no longer measured and is 6/7 of the maximum current as given by Equation (1).

There are two advantages in using a constant preselected drop time in polarography. The first is that with a normal, free-falling drop, t—and hence i_d—will vary as a function of potential [12]. This may cause problems in analysis, especially at very negative potentials. The second advantage is the ability to decrease the time of analysis. For a naturally falling drop of 5 seconds the potential sweep rate will usually be about 2 mV/sec. Thus for a scan range of 1.5 volt this would mean an analysis time of 750 seconds. However, by using a drop time of 0.5 to 1.0 second, the scan rate may be increased to between 10 and 20 mV/sec and the time of analysis reduced from

750 seconds to between 75 to 150 seconds. The only important point for analysis is that there be sufficient drops on the rising portion of the wave to define the shape of the current-voltage curve and to determine $E_{1/2}$.

For a series of octaethylporphyrins, the average diffusion coefficient, D, was measured as $1-3 \times 10^{-6}$ cm^2/sec in DMSO, DMF, or butyronitrile, and is almost identical for the oxidized and reduced forms of the complex.* Similar values are obtained for iron complexes of tetraphenylporphyrin* and complexes of tetraphenylporphyrins substituted at the β-position on the porphyrin ring [14]. Thus, assuming values of $n = 1$, $t = 1$ sec, and $m = 2$ mg/sec, one could expect a value of $1-2$ μA for 1.0 mM porphyrin in DMSO. As the solvent is changed, $D^{1/2}$, and hence the current, will vary inversely with the square root of the solution viscosity [12].

Analysis of the wave shape on the rising portion of the wave according to Equation (2) is called wave analysis. We have

$$E_{\text{DME}} = E_{1/2} - \frac{0.059}{n} \log \frac{i}{i_d - i} \quad \text{at } 25°\text{C}, \quad (2)$$

where E_{DME} is the potential of the DME and $E_{1/2}$ is the half-wave potential measured when the current is one-half that of the diffusion-limited plateau. We also have an equation relating $E_{1/2}$ to the formal potential, $E^{0\prime}$:

$$E_{1/2} = E^{0\prime} + \frac{0.059}{n} \log \frac{D_R^{1/2}}{D_O^{1/2}} \quad \text{at } 25°\text{C}, \quad (3)$$

where D_O and D_R are the diffusion coefficients of the oxidized and reduced species. Since these have been measured as being almost identical for most complexes of metalloporphyrins, $E_{1/2}$ is virtually identical to $E^{0\prime}$ for reversible reactions. According to Equation (2), plots of E_{DME} versus $\log[i/(i_d - i)]$ should give a straight line with slope $-0.059/n$ and yield directly the value $E_{\text{DME}} = E_{1/2}$ at $\log[i/(i_d - i)] = 0$.

If the electrode reaction is reversible (rate of electron transfer $\gg$ rate of mass transport), a theoretical slope of -0.059 volts will be obtained for a one-electron transfer and the measured $E_{1/2}$ may be used as a thermodynamic quantity similar to $E^{0\prime}$. This technique was initially utilized by Clack and Hush [8] to confirm the one-electron-transfer steps of metalloporphyrins. However, if the rate of electron transfer is not rapid, the measured $E_{1/2}$ consists of both a thermodynamic and a kinetic term. The wave appears to be drawn out, and in this case wave analysis gives a slope greater than $-0.059/n$ volt or an apparent $n < 1$. The second problem is that the measured $E_{1/2}$ will be displaced in a negative (cathodic) direction along the potential axis, with the

---

* Kadish, K. M., and Davis, D. G., unpublished results.

magnitude of the displacement depending upon the electron-transfer rate constant k^0. In this case there is no thermodynamic significance to the $E_{1/2}$ measured at the DME, and care must be taken not to assume one.

ii. Cyclic Voltammetry

Cyclic voltammetry is now the most widely used method for measuring redox potentials of metalloporphyrins. The advantages of this technique are numerous when compared to the earlier techniques of potentiometry and polarography. First, the potential of a redox couple may be rapidly determined from the current-voltage curve. Secondly, the chemical stability of the electrogenerated species may be ascertained by reversing the potential sweep and observing the electrode reactions of the product. Furthermore, in the case of coupled chemical reactions, a variation of the potential sweep rate and measurement of the new current-voltage curve might enable determination of the type of coupled chemical reaction and the mechanism of electron transfer. This has been discussed in detail by Nicholson and Shain [15], as well as Saveant and others (for a good review see [16]). Finally, the standard heterogeneous electron-transfer rate constant may be estimated from the difference in anodic and cathodic peak potentials as a function of scan rate [17]. Each of these points will be illustrated in the following pages. In discussing the technique of cyclic voltammetry we will consider first the theory for reactions in which there is a fast (reversible) electron transfer, and then that for which there is a quasireversible charge transfer. In the latter case rate constants of electron transfer may be determined directly from the current-voltage curves. The cases of irreversible electron transfer (which gives no reverse peak [15]) and coupled chemical reactions associated with the electron-transfer step (which also may give current-voltage curves without a reverse peak) will not be discussed in this review, since they have not been dominant in the electrochemistry of iron porphyrins studied to date.

Both solid and mercury electrodes may be used in cyclic voltammetry. For the case of mercury, a hanging mercury drop electrode (HMDE) may be employed. For the case of solid electrodes, the electrode material most often used in porphyrin research has been platinum.

Four sets of peaks are observed in Figure 2. Reactions (3) and (4) correspond to the successive reduction of TPPFeCl and TPPFe, while reactions (2) and (1) corresponds to the successive oxidation of TPPFeCl by single-electron-transfer steps. The illustration shown is an example of a cyclic voltammogram obtained by multisweep techniques, that is, the potential was varied continuously between $+1.60$ and -1.30 V versus SCE. This is usually not the best way to obtain data of this sort and is used mainly for illustrative purposes.

The concentration of electroactive species utilized for cyclic voltammetry may be more dilute than that utilized in polarography, due to the sensitivity of the technique. Thus, one may use 10^{-4} M or less. However, the higher the

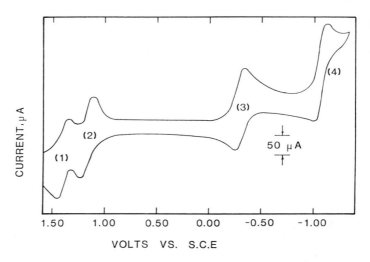

Figure 2 Cyclic voltammogram of TPPFeCl in CH_2Cl_2, 0.1 M TBAP. Platinum electrode. (Reprinted with permission of American Chemical Society [50].)

concentration (up to about 5×10^{-3} M), the better looking the curve and the easier it is to measure peak potentials. Because of this, most of the data in the literature have been taken with millimolar concentrations of porphyrins. As in polarography (and all electrochemical techniques), a supporting electrolyte must be employed, which is usually at least 0.1 M in concentration.

For the oxidation-reduction of TPPFeCl the potential should be reversibly scanned, first in one direction (for example to observe reductions) and then in another direction (in this case to observe oxidations). Thus one might scan initially from $+0.2$ to -1.3 V and back to $+0.2$ V to record reactions (3) and (4), and then, in a second experiment, use the same solution from 0.0 to $+1.6$ V to observe reactions (2) and (1). This starting potential (0.2 V) would be selected because the species in solution, TPPFeCl, is nonelectroactive at 0.2 V and is neither oxidized nor reduced.

It is not necessary to investigate all of the peaks at the same time. For example, to investigate only the reaction Fe(III)/Fe(II), the potential is scanned first in a cathodic direction through the standard potential of the Fe(III)/Fe(II) couple (in this case -0.29 V), then reversed at some potential between -0.5 and -0.9 V in order to investigate the backward reaction. If there are no chemical complications, as in this case, the forward sweep corresponds to the reaction TPPFeCl → TPPFe and the backward sweep to TPPFe → TPPFeCl.

There are several diagnostic criteria which must be considered when analyzing current-voltage curves. These include analysis of both the shape and the position of the peak as well as the maximum peak current with respect to

Techniques

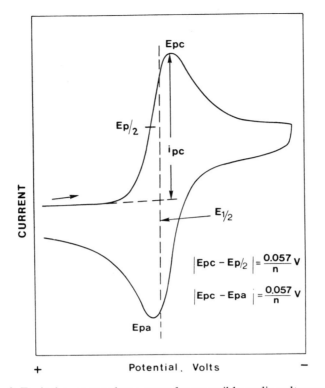

Figure 3 Typical current-voltage curve for reversible cyclic voltammogram.

scan rate. For a reversible reaction in which there are no chemical complications, the shape and position of the wave remain independent of the sweep rate. Thus, as shown in Figure 3, the cathodic peak potential $E_{p,c}$ and the anodic peak potential $E_{p,a}$ remain constant when the scan changes.

Again assuming that the electroreduction is reversible, the following relationships may be observed:

$$E_{p,c} - E_{1/2} = -0.0285/n \text{ volts}, \qquad (4a)$$

$$E_{p,a} - E_{1/2} = 0.0285/n \text{ volts}, \qquad (4b)$$

$$E_p - E_{p/2} = -0.057/n \text{ volts} \qquad \text{at } 25°C. \qquad (5)$$

Combining Equations (4a) and (4b) yields

$$E_{p,a} - E_{p,c} = 0.057/n \text{ volts} \qquad (6)$$

Values of n, the number of electrons transferred, may be calculated from

Equations (4a), (4b), (5), or (6), and should agree within 20%. Deviation greater than this often implies separate reaction mechanisms in oxidation and reduction.

In addition, $E_{1/2}$, which is midway between $E_{p,c}$ and $E_{p,a}$, may be calculated from the algebraic addition of Equations (4a) and (4b):

$$E_{1/2} = \frac{E_{p,c} + E_{p,a}}{2}. \tag{7}$$

In principle, Equation (7) may always be applied even if the rate of electron transfer is slow and the separation is greater than $59/n$ mV. This will be true as long as an oxidation and a reduction peak are obtained which are of approximately equal height and correspond to the reaction Fe(III) ⇌ Fe(II).

It might be mentioned that in practice one may observe deviation from Equations (4)–(6) due either to quasireversible electron transfer or to uncompensated *IR* loss in solution. This latter effect is most often observed at high potential sweep rates but may also occur when there is too high an ohmic resistance in the bridge and the reference electrode or too low a concentration of supporting electrolyte.

In order to ascertain the reversibility of a given redox couple the potential sweep should be varied and the constancy of E_p verified. In addition, the peak current i_p should increase with the square root of the scan rate v. The use of the quantity $i_p/v^{1/2}$ comes from an equation relating peak current to sweep rate. This equation, called the Randles-Sevcik equation, is given as follows:

$$i_p = 2.69 \times 10^5 n^{3/2} A D_0^{1/2} v^{1/2} C, \tag{8}$$

where i_p is in amperes, the electrode area A in cm^2, D_0 in cm^2/sec, C in mol/cm^3, and v in V/sec. Constancy of $i_p/v^{1/2}$ indicates that at the scan rate investigated (that is, for the effective time of the experiment) there are no coupled chemical reactions. Variation of this quantity with scan rate can be used to prove nonreversible behavior and to probe the type of reaction mechanism which may occur [15] in the time scale of the experiment.

Finally, in considering reversible reactions the ratio of anodic to cathodic peak currents may be utilized. If the diffusion coefficients of the oxidized and reduced species are equal, it is expected that the anodic current maximum i_{pa} will be equal to the cathodic current maximum i_{pc} and the ratio $i_{pa}/i_{pc} = 1$. This, although true, may be difficult to measure if there are close, overlapping reactions or the peak occurs on the rising portion of the background wave. The former is illustrated by reactions (1) and (2) and the latter by reaction (4) of Figure 2.

Cyclic voltammetry may also be used to measure the heterogeneous electron-transfer rate constants associated with the reversible or quasireversible reaction Ox ⇌ Red. The method employed is quite simple, and values of k^0 can be obtained at the same time as measurements of $E_{1/2}$ are taken [17].

Techniques

In this technique, peak potentials are measured for the redox couple of interest as a function of potential sweep rate. For rapid reactions the potential difference ($E_{pa} - E_{pc}$) will be equal to $0.057/n$ volts [see Equation (6)], but this will depend on the scan rate. As the scan rate is increased, a point can be reached where a shifting of both the anodic and the cathodic peak from $E_{1/2}$ will occur, and it is in this range that one can measure k^0. Care must be taken that all residual IR losses are corrected for or eliminated. This is accomplished by both use of a Luggin capillary and an electronic technique called positive feedback. Most modern instruments have positive feedback already built into the instrumentation.

At the point where shifts from reversibility are observed, the rate of electron transfer can be calculated using the Nicholson equation [17] for a quasireversible system:

$$k^0 = \psi(\gamma)^{-\alpha}\left(\frac{\pi n F}{RT}D\right)^{1/2} v^{1/2}, \tag{9}$$

where k^0 is the standard heterogeneous electron-transfer rate constant at the standard potential E^0, n is the number of electrons transferred, γ is the ratio of the diffusion coefficients of the oxidized and reduced complex, D_{ox}/D_{red}, α is the transfer coefficient (usually around 0.5), and v is the potential scan rate. The most important factor in this equation is ψ, which is related to ΔE_p. Computer-generated values of ψ for each ΔE_p (actually $\Delta E_p \times n$) have been tabulated after solution of the appropriate numerical equations [17] and are reproduced in Table 1.

The strength of the cyclic voltammetry technique is that values of scan rate can be varied over a large range of magnitude in order to find the appropriate values of $n \Delta E_p$. Constancy of k^0 calculated at several scan rates shows that the mechanism (simple electron transfer without complications) is probably correct.

Finally, in looking at Equation (9), several simplifying assumptions can be made, especially for the case of metalloporphyrins. In almost all solvents

Table 1. Variation of Peak Potential Separations with Kinetic Parameter ψ for Cyclic Voltammetry[a]

$n \Delta E_P$ (mV)	ψ	$n \Delta E_P$ (mV)	ψ
61	20	84	1
63	7	92	0.75
64	6	105	0.5
65	5	121	0.35
66	4	141	0.25
68	3	212	0.1
72	2		

[a] Data taken form R. S. Nicholson, Reference [17].

$D_{ox} \simeq D_{red}$, so that $\gamma \simeq 1$. Therefore $\gamma^{-\alpha}$ is also approximately unity and can be dropped from the equation. The number of electrons transferred, n, is always one for the reaction Fe(III) $\rightleftarrows$ Fe(II), and the value of $F/RT = 39.2$ V^{-1} at 25°C. For most iron porphyrins the measured diffusion coefficients are approximately 2×10^{-6} cm^2/sec. Thus, the equation may be simplified so that k^0 is directly related to ψ by the scan rate:

$$k^0 \simeq 0.0157 \psi v^{1/2}. \tag{10}$$

An example of rate-constant determinations is given in Table 2. For the values listed in this table an exact solution using Equation (9) was obtained. Equation (10) is presented only for illustrative purposes and should not be utilized directly.

When attempting to measure heterogeneous electron-transfer rate constants by cyclic voltammetry it is of interest to know what range of values may be determined by the technique of Nicholson and also how k^0 compares with values of rate constants measured for homogeneous electron transfer of the same redox couple. Both of these questions can be answered by simple calculations.

The range of potential sweep rates normally utilized for rate measurement is between 0.010 V/sec and 100 V/sec in cyclic voltammetry. Given a peak separation $|E_{p,a} - E_{p,c}|$ of 61–212 mV for a one-electron transfer, we can calculate with the aid of Equation (10) and Table 2 what range of k^0's might be measured using the technique of cyclic voltammetry. Using the extreme values of $\psi = 0.1$ and $\psi = 7$ it is seen that heterogeneous electron-transfer rate constants can be measured in the range of 1 cm/sec at the fast end to 1.6×10^{-4} cm/sec at the slow end. Outside of these limits other techniques must be employed.

One of the problems facing porphyrin electrochemists is how to transpose electrochemical results of porphyrin redox reactions into a language understandable to inorganic chemists and biochemists. There is usually no problem

Table 2. Calculation of Heterogenous Electron-Transfer Rate Constant for Reduction of EtioFeCl in DMF, 0.1 M TEAP[a,b]

Scan Rate v (V/sec)	$v^{1/2}$	ΔE_p (mV)	ψ	k^0 (10^{-4} cm/sec)
0.062	0.249	175	0.18	8.4
0.094	0.307	181	0.17	9.8
0.125	0.354	195	0.14	9.3
0.157	0.396	202	0.12	9.0
0.188	0.434	208	0.11	9.0
0.220	0.469	215	0.094	8.3
			Avg =	9.0 ± 0.3

[a] Data taken from K. M. Kadish and G. Larson, Reference [140].
[b] Calculated from Equation (9) using $D_{ox} = D_{red} = 2.9 \times 10^{-6}$ cm/sec.

in this respect in reporting half wave or peak potentials, since these values are clearly related to the standard potential E^0. This is not the case, however, with an electrochemical rate constant. For one thing, the rate constant is potential-dependent and increases or decreases exponentially as a function of $E - E^0$. This is unlike all chemical rate constants, which are potential-independent. Secondly, the dimensions of the rate constant, cm/sec, are not readily converted to those of chemical concentrations and conventional rate constants. Several correlations between chemical and electrochemical rate constants have been provided by the Marcus theory [18]. One such relationship relates the standard heterogeneous rate constant k^0 and the rate constant k_{11} for homogenous electron transfer.

The standard rate constant k^0 can be represented by the rate constant for the reaction (11) below at E^0, while the self exchange is represented by the reaction (12):

$$Fe(III) \rightleftharpoons Fe(II), \qquad (11)$$

$$Fe(II) + Fe^*(III) \rightleftharpoons Fe(III) + Fe^*(II). \qquad (12)$$

The values of k^0 and k_{11} are related by

$$k^0 = Z_{el} \left[\frac{k_{11}}{Z_{sol}} \right]^{1/2} = \left[\frac{k_{11}}{10^3} \right]^{1/2}, \qquad (13)$$

where Z_{sol} and Z_{el} are the collision frequencies of the chemical and electrochemical reactions, respectively (10^{11} liter mole^{-1} sec^{-1} and 10^4 cm sec^{-1}). Although there are a number of assumptions associated with the use of this equation (the most important is that all reactions are outer-sphere), it appears to work in many cases.

iii. Coulometry and Controlled-Potential Electrolysis

A combination of coulometry and controlled-potential electrolysis produces one of the more important techniques that may be used in the study of metalloporphyrin redox reactions. Controlled-potential electrolysis may be accomplished in a variety of different cells and allows for the complete generation of a new electrochemical species from the starting material. Electrochemical reactivity, as well as the spectral properties and ESR characteristics of the new complex, may be investigated after complete generation *in situ* if the products are either long-lived or stable. Because the potential can be easily varied, a highly selective oxidant or reductant is available which can selectively generate new compounds by single electron additions or abstractions in the absence of added chemical interferences. In addition, controlled-potential coulometry can often be used to generate novel species which, in the presence of suitable anionic or cationic species, may be easily isolated for identification in the solid state.

In addition to its use for selective generation of novel species, controlled-potential coulometry can be used to determine the overall number of electrons transferred in an electrode reaction. Although the techniques of polarography and cyclic voltammetry give a good idea of the number of electrons transferred in the oxidation reduction step, it is important to confirm this by exhaustive controlled-potential electrolysis. In many cases the value of n may be different on the short time scale of cyclic voltammetry and on the longer time scale of controlled-potential electrolysis. A good example of this is provided by the data of Kadish and Jordan for the electroreduction of iron protoporphyrin IX dimers in aqueous media [19]. Using rapid-scan cyclic voltammetry, a 60-mV peak separation was obtained and $n = 1$ calculated for the reduction. Controlled-potential electrolysis yielded an overall $n = 2e$ per dimer unit (or $1e$ per monomeric iron). Based on this combination of data, an overall reduction mechanism involving two one-electron-transfer steps could be formulated and the presence of an intermediate mixed oxidation state dimer confirmed. Another example of coulometry and controlled-potential electrolysis being essential to mechanism determination is provided by Kadish and coworkers for the reduction of (TPPFe)$_2$O in DMF [20]. The first reduction peak of this complex was irreversible and consisted of several overlapping processes, and so little information concerning n could be obtained by cyclic voltammetry. However, controlled-potential electrolysis when combined with spectroscopic and ESR data gave evidence for the overall reduction of one (TPPFe)$_2$O dimer to yield two Fe(I) monomers. The various advantages of coulometry as an electroanalytical method have been discussed in the literature [21, 22]. Basically, in this technique all of the electroactive species is reduced and the total quantity of electricity determined. The current for this reaction decreases as a function of time, and the integrated current Q is measured during this total electrolysis. The desired final quantity, Q_F, is the difference between the total charge Q_T and the background charge Q_B and is related to the concentration by Faraday's law:

$$Q_F = Q_T - Q_B = nFN^0 = nFVC, \qquad (14)$$

where N^0 represents the total number of equivalents initially present, n is the number of electrons transferred per molecule, F is Faraday's number (96, 487 C eq^{-1}), V is the solution volume, and C is the concentration of the electroactive species.

Solid electrodes must be used for oxidations, while either Pt or Hg electrodes may be used for reductions. A number of cell designs are possible, and these are discussed in the literature [13, 21, 22]. Of utmost importance is that the working and the reference electrode be physically separated and that the frit utilized have as small a leakage as possible. The time of electrolysis should be as short as possible, so small volumes and efficient stirring are a prerequisite. The concentration of porphyrin utilized is unimportant, since only the total number of coulombs is recorded. In practice, however, a concentration should be selected which is close to that of the electrochemical technique utilized (10^{-3} to 10^{-4} M). In this manner any unusual reactions, such as polymerization or precipitation, can be observed under the same experimental conditions as those for which the current-voltage curves were recorded. The potential for electrolysis should be at least 100 mV beyond $E_{1/2}$ if possible. The larger the difference between the applied potential and $E_{1/2}$, the faster the reaction will proceed, provided that stirring is not the rate-determining step. Care should be taken, however, not to electrolyze on the rising part of the solvent ramp, since this generates a number of side products which will invariably react with the compound of interest.

Generally the time of electrolysis is between 15 and 30 minutes for complete electrolysis. This will depend on the exact electrode configuration and the concentration of the porphyrin in solution. More rapid electrolysis may be obtained in thin-layer cells where complete oxidation or reduction can be accomplished in 25–60 seconds. Coulometric measurements can also be made in the thin-layer cell with a knowledge of the cell volume, which will vary between 0.2–50 μl, and is usually electrochemically calibrated.

iv. Joint Application of Electrochemical and ESR Techniques

Electrochemistry may serve as an elegant tool for the generation of free radicals in solution and their investigation by ESR. Since electrode potentials may be easily controlled, a given radical can be cleanly generated in solution and studied in the absence of chemical oxidants of reductants.

Numerous π-cation and π-anion radicals of porphyrins have been generated by electrochemical techniques and investigated by ESR [23]. In most cases the technique may be used to differentiate metal from ring-centered reactions, as in the reduction of Fe(II) porphyrins [9]. In addition, mixed-oxidation-state iron dimers have also been characterized by ESR after electrochemical generation [20].

In many cases electrochemical generation of a paramagnetic species may be done outside the ESR cavity and a transfer made after electrooxidation or electroreduction. This technique is not recommended; when possible, electrochemical generation of the diamagnetic species should be done directly in the ESR cavity. Several cells are commercially available. Others can be made to suit the specifications of the individual laboratory. One such cell is shown in Figure 4(a).

Both the temperature of the solution and the solvent system may be varied in order to obtain a stable generated species. The presence of tetrabutylammonium salts does not generally interfere with the ESR measurement, since these cations are not known to form ion pairs. In many cases π-anion and

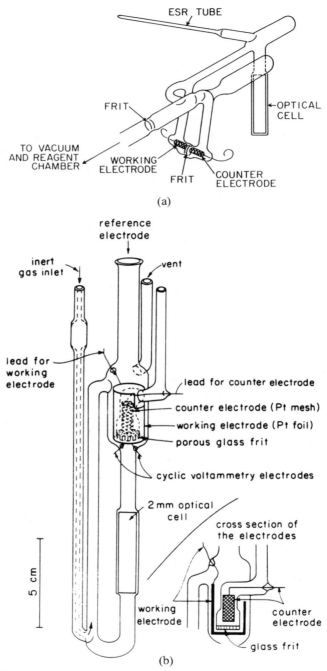

Figure 4 (a) Electrochemical cell for ESR-optical measurements *in vacuo*; (b) electrooptical cell for controlled potential electrolysis, cyclic voltammetry, and optical measurements; and (c) electrochemical cell for *in vacuo* electrolysis. (Reprinted with permission of the American Chemical Society [73].)

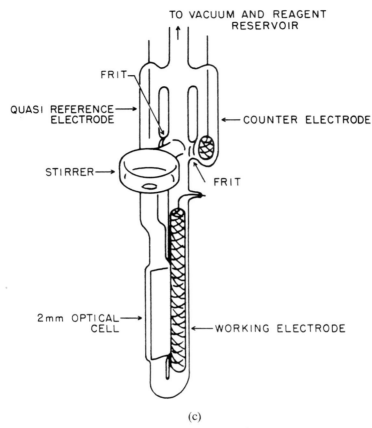

(c)

Figure 4 *Continued*

π-cation radicals of metalloporphyrins have been shown to be stable for hours. Selection of a good solvent for radical stability is discussed in the literature [13, 24, 25].

The data which may be obtained from the combined electrochemical-ESR technique includes all of the usual ESR characteristics such as the g-factors and the hyperfine coupling constants. These values are in turn related to the structural state of the radical or metal ion and will give extensive information regarding the mechanism of electroreduction or electrooxidation. When they are used in combination with the spectroelectochemical techniques, a better characterization of the solution properties and metal oxidation state of the porphyrin may be obtained. In the case of iron porphyrins the stable oxidation state is Fe(III), which is paramagnetic and has been well characterized in the literature. Unfortunately, Fe(II) is diamagnetic and so no ESR data exist on this important oxidation state of the complex. Starting with Fe(III) complexes, definitive ESR assignments have only been obtained for electrochemically generated $[PFe(I)]^-$ [9] and $[(PFe)_2O]^-$ [21]. In coming years much more information may become available on the oxidation or reduction products of a number of novel dimeric iron porphyrin complexes whose synthesis and physical properties are now just being described in the literature.

v. Spectroelectrochemistry

A combination of spectroscopic and electrochemical techniques is invaluable for identifying the products and electron-transfer mechanisms of porphyrin electrooxidation or electroreduction. The simplest approach is to transfer a portion of the solution to a spectrophotometric cell after complete controlled potential electrolysis. Care in excluding oxygen is absolutely necessary for these operations. In some cases, reasonable spectra of reduced or oxidized species may be obtained. The problem, however, is to know if the observed spectrum is the product of the electrode reaction or is a new species produced during transfer from the electrochemical cell to the spectrophotometer. A surer approach which has been utilized by several laboratories is to make spectroscopic measurement directly in a cell in which controlled-potential electrolysis has been carried out. These may be large coulometric cells in which an optical window has been inserted, as in the case of a cell designed by Fajer et al. [26] [see also Figure 4(b), (c)], or they may be small-volume optically transparent thin-layer cells [27].

A large coulometric cell with optical windows has been used to determine the spectra of a number of oxidized metalloporphyrins, including those of oxidized TPPFeCl and $(TPPFe)_2O$ in CH_2Cl_2 [5, 6]. A similar design which fits directly into the spectrometer was used by Lexa and coworkers to spectrally identify the electrode products of reduced TPPFeCl and $(TPPFe)_2O$ in DMF [9]. The advantage of a large coulometric cell is simplicity of operation and the low cost involved in the spectroscopy. The disadvantage is that a large volume of solution is required, electrolysis is slow (15–30 minutes or longer), and the measurement is limited to species stable for 15 minutes to an hour or longer. This luxury of large volumes and species stability is not always available. Furthermore, with the use of these cells reaction intermediates are difficult to observe and determination of spectral reversibility can be difficult. For this reason, a number of determinations are now being made with optically transparent thin-layer electrodes (OTTLE). A review on thin-layer electrochemistry appeared several years ago [28]; a good example of the method as applied to metalloporphyrins has been given by the laboratories of Wilson [29–31] and of Kadish [32]. Suitable equations for analyzing spectral data have also been derived by Ryan and Wilson [33].

Thin-layer cells are ideally suited for studying spectral reaction intermediates in the oxidation or reduction of metalloporphyrins. The path length of the cell is usually less than 0.03 cm, and the cell volume can be as small as 0.2–50 µl. This has the dual advantage of a rapid electrolysis time (30–60 seconds) and the utilization of only a small amount of porphyrin, which may be costly or difficult to obtain. For the obtaining of spectral intermediates rapid-scanning spectrometers can be used, which are able to acquire a complete spectral region in 5–100 μs.

Many advances have been made in OTTLE cell design since the initial report of the glass-slide–gold-minigrid assembly [27] for spectroelectrochemistry. These include the capability for monitoring in the ultraviolet and infrared spectral regions [34], the ability to perform precise [35] and cryogenic [36] temperature studies, and the flexibility of several working electrode materials [36–39]. In addition, cells have been specifically designed to allow for ease of solution deoxygenation [40] and for reduction of uncompensated solution resistance [41].

Historically, most thin-layer studies have been restricted to aqueous solutions and to nonaqueous solvents like dimethylformamide and acetonitrile. While these solvents are extremely useful, they are not ideal for porphyrin studies from either a solubility or a mechanistic point of view. With this in mind a new cell design has been reported which is both easy to construct and inert to solvents of a wide range of dielectric constant and binding properties [42].

C. SOLVENT SYSTEMS AND SUPPORTING ELECTROLYTES

i. Aqueous Solutions

Most early studies of iron porphyrins were carried out in aqueous media or in mixed ethanol-water mixtures. Solutions were buffered to the desired pH with the use of appropriate buffer systems. Usually the supporting electrolyte was the buffer itself and no additional supporting electrolyte needed to be added. The potential range of an aqueous solution will depend on the type of electrode material and the pH of the solution. In an aqueous 0.1 M KCl solution the potential window of a Hg electrode (either a DME or HMDE) will range between -1.9 V versus SCE on the cathodic side and 0.0 V on the anodic side. The negative limit is determined by reduction of the potassium ion. The anodic limit is dependent upon the dissolution of mercury, which is facilitated in the presence of anions such as halides, cyanide, hydroxides, thiosulfates, or thiocyanates which precipitate or complex Hg(I) or Hg(II) ions. Changing the solvent from 0.1 M KCl to either an acidic or a basic solution will drastically shift the cathodic limit of the solvent. Since the limit is determined by the reduction of H^+ ions, basic solutions will have a wider cathodic range than acidic solutions. For example, in 1 M TEAH a limit of almost -2.5 V may be obtained, while with 1 M HClO$_4$ barely -1.0 V versus SCE is obtained.

A greatly decreased cathodic-potential limit is observed on platinum or carbon electrodes. Similarly to Hg, the cathodic limit is determined by reduction of H^+ or reduction of the cation from the supporting electrolyte. Cathodic limits on Pt range from approximately -1.0 V in 1 M NaOH to -0.2 V in 1 M HClO$_4$. These limits should not be considered absolute and are presented only as general guidelines to the solvent-system range.

Anodic limits on platinum or carbon will be determined by the oxidation of OH$^-$ and will also be dependent on pH. For oxidations on platinum this potential will range from $+0.4$ V in 1 M NaOH to almost 1.3 V in 1 M HClO$_4$. Again these values should be taken only as general indications of the solvent range. Reduction of H$^+$ or oxidation of OH$^-$ will occur over a wide potential range, and the definition of a "limit" is an operational one depending upon the given current for solvent oxidation-reduction.

Because early studies of iron-porphyrin electrochemistry were concerned only with the reactions of Fe(III) and Fe(II), there were few problems (other than solubility) in using aqueous solutions. For all ligand and porphyrin systems investigated the potential for Fe(III)/Fe(II) in H$_2$O or even H$_2$O-ethanol systems was well within the range of a platinum electrode. Thus, potentiometric measurements could easily be made. Changes of pH in order to obtain solubility or to study the OH$^-$/H$_2$O ligand exchange also presented no problem. In addition, early thinking was that water was the biological medium and that there was little reason to extend studies to other, "nonbiological", aprotic solutions.

However, during recent years, there has been some discussion regarding water versus aprotic media for studies of metalloporphyrin redox reactions. In some cases it may be argued that the biological medium more closely approximates an aprotic solvent than a protic solvent, and so studies in aprotic media are "justified." In addition, use of aprotic media allows for a much expanded potential range with respect to that obtained with aqueous media.

ii. Aprotic Solvents

In recent years a large number of electrochemical studies have been carried out in aprotic media. For investigations of porphyrins early studies were mainly in DMF or DMSO for reductions, and in benzonitrile or butyronitrile for oxidations. Later studies of porphyrin electrochemistry have made use of the nonbinding solvent CH$_2$Cl$_2$ for both oxidations and reductions. The advantage of this solvent is that both radical anion and radical cations may be reversibly formed in solution and investigated by the techniques of ESR and NMR as well as by u.v.-visible absorption spectroscopy. For organic electrochemistry—that is, the electrochemistry of the conjugated π-system—aprotic solvents are the solvents of choice. The selection of which aprotic solvent to utilize may depend, in addition to its potential range, on its ability to stabilize anion or cation radicals. For example the stability of anion radicals is greater in DMF and DMSO than in acetonitrile. This may be due to the fact that solvents with a large donor number tend to associate with hydrogen donors such as water. Thus, despite the presence of small amounts of water in DMSO or DMF, stable anion radicals may be obtained.

Table 3. Physical Properties and Maximum Potential Range of Several Common Solvents used for Porphyrin Electrochemistry

Solvent	(abbreviation)	F. P. (°C)	B. P. (°C)	Cathodic limit[a]		Anodic limit[a]	
				Pt	Hg	Pt	Hg
1,2-dichloroethane	(EtCl$_2$)	−35	83.4	−1.9	—	+1.8	—
Methylene chloride	(CH$_2$Cl$_2$)	−96.7	40	−1.9	−1.9	+1.8	0.6
Benzonitrile	(BN)	−13.5	191.1	−1.8	—	+1.7	—
Acetonitrile	(AN)	−45.7	81.6	−2.0	−2.8	+1.8	+0.6
Tetrahydrofuran	(THF)	−108.5	65	−3.6	−3.6	+1.8	0.0
Dimethylformamide	(DMF)	−61	153	−2.5	−3.0	+1.6	+0.5
Dimethylacetamide	(DMA)	−20	165	—	—	+1.3	—
Dimethylsulfoxide	(DMSO)	18.5	189.0	−1.9	−3.0	+0.7	+0.4
Pyridine	(py)	−41.6	115.5	−1.8	−1.6	+0.8	—

[a] Values given are the maximum limits observed in the literature and are reported versus SCE. The observed values may be substantially less, depending on the type of supporting electrolyte and the purity of the solvent.

The purification procedure, potential limits, and physical characteristics of numerous nonaqueous solvents have been given in the literature [13, 24, 25]. Table 3 reproduces some of these data for the solvents most often used in the electrochemistry of iron porphyrins. The abbreviation of the solvent, the boiling and freezing point of the solvent, and the anodic and cathodic limits are given in this table.

Selection of a given solvent will depend on the requirements of the individual experiment. The first requirement, of course, is solubility. One must also consider ease of purification, the ability of the solvent to stabilize anion or cation radicals, and the potential range of the solvent. This will in turn depend on the type of electrode used. Other practical factors include the cost of the solvent, its toxicity, and the general ease of handling it.

As already mentioned, the most common nonaqueous solvent utilized for iron porphyrin reductions has been DMF, with DMSO a close second. With appropriate supporting electrolyte, the range of these solvents may be extended to −3.0 V using a Hg electrode. This easily allows measurement of anion radicals and dianions as well as any following electrode reactions involving protonated species which are observed in the range of −2.2 to −2.7 V [29–31]. Anion radicals are easily stabilized in this solvent for investigation by ESR techniques.

The boiling points of DMF and DMSO as well as the freezing point of DMF are also ideal for electrochemical measurements of iron-porphyrin reactions. In contrast with low-boiling-point solvents such as CH$_2$Cl$_2$, there is little evaporation during the experiment, and a constant concentration may be obtained for long periods of time. This is especially important for experiments done in thin-layer cells. In this case the initial cell volume will be microliters, and changes in solution volume due to evaporation must be strictly avoided.

DMF is ideal as a solvent for low temperature, and for variable-temperature experiments with iron porphyrins DMF is ideal. Kadish and Schaeper have investigated reductions in this solvent at temperatures between 225 and 300 K [43]. Because of the relatively high dielectric constant of DMF, few problems in *IR* loss were encountered at low temperatures. This is not the case in CH_2Cl_2, where the resistance increases substantially as the temperature is lowered.

Most oxidations of porphyrins have been carried out in CH_2Cl_2, benzonitrile, or butyronitrile. Although the anodic-potential range of these solvents does not extend much above +1.5 V, this is sufficient to view the first two oxidations of most monomeric iron complexes, which occur in the range of +0.9 to 1.3 V. Since cation radicals and dications are attacked by nucleophiles to yield an isoporphyrin [5, 44], the solvent of choice is CH_2Cl_2. This solvent has been utilized by numerous laboratories, especially for characterization of the first oxidation of Fe(III) complexes. There is also an added advantage in that comparisons may be made directly between electrochemical results obtained in this solvent and chemical results obtained in other nonbonding solvents such as CCl_4 or $CHCl_3$.

One disadvantage of CH_2Cl_2 is its low boiling point. Depending on the cell design, this may present special problems in normal cells upon deoxygenation as well as in thin-layer spectroelectrochemistry, where time-resolved measurements over a long time are difficult due to solvent evaporation under the influence of a light source. In order to get around this problem one can easily switch to the solvent 1,2-dichloroethane in place of dichloromethane. Because of its higher boiling point (83.4°C), fewer problems are encountered due to solvent evaporation. The dielectric constant of the solvent is 10.4, compared to 8.93 for CH_2Cl_2, and its donor number is defined as 0.0, similar to CH_2Cl_2. The electrochemical mechanisms and ligand-binding properties of iron porphyrins appear to be identical in these two solvents, so that results can be used interchangeably.* There are, however, slight differences in absolute potential measurements between the two solvents. These exist even after correction for liquid junction potential and may be due in part to the different dielectric constants of the two solvents.

iii. Supporting Electrolytes

Almost all electrochemical studies of porphyrins in nonaqueous media have utilized tetraalkylammonium salts as supporting electrolyte. The most common have been the tetrabutyl- and tetraethylammonium perchlorates (abbreviated TBAP and TEAP). Several studies have also utilized the tetraalkylammonium salts of BF_4^- and PF_6^-.

---

* Kadish, K. M., unpublished observations.

Table 4. Accessible Potential Range (V versus SCE) of Several Supporting Electrolytes in CH_2Cl_2[a]

Supporting electrolyte	Electrode material	
	Platinum	Mercury
Tetrabutylammonium perchlorate	1.8 to -1.7	0.8 to -1.9
Tetrabutylammonium chloride	0.86 to -1.9	-0.35 to -1.7
Tetrabutylammonium bromide	0.59 to -1.7	-0.37 to -1.7
Tetrabutylammonium iodide	0.18 to -1.7	-0.45 to -1.7

[a] Data taken from C. K. Mann, "Nonaqueous Solvents for Electrochemical Use," in *Electroanalytical Chemistry*, A. J. Bard, (ed.), New York, 1969, Vol. 3, p. 123.

Selection of a supporting electrolyte will depend upon the cost, ease of purification, and binding of the counteranion. Usually ClO_4^-, BF_4^-, and PF_6^- are considered as nonbinding anions. Tetraalkylammonium supporting electrolytes containing all three anions are commercially available or may be easily synthesized from the respective Br^- complexes [45]. However, the ClO_4^- salt with either the tetraethylammonium or the tetra-n-butylammonium counterion has usually been the supporting electrolyte of choice. Results obtained between the two salts are almost interchangeable, and thus given the choice (which depends on availability) TBAP has usually been selected because of its higher solubility. For reactions of iron porphyrins $TBA^+ PF_6^-$ should never be used. Dissociation of PF_6^- to yield PF_5 and F^- is known to occur, with a resulting complexation of F^- by both Fe(II) and Fe(III). Thus, use of a seemingly nonbinding supporting electrolyte at 0.1 to 1 M concentration may lead to mM concentrations of F^- ion which will form five- and six-coordinated complexes with the iron [46].

Table 4 presents data on the potential range in CH_2Cl_2 of several supporting electrolytes. As seen from this table, salts of I^-, Br^-, or Cl^- may be used for reductions, but may present problems with oxidations due to the fact that they are easily oxidized to I_3^-, Br_3^-, and Cl_2 at potentials less than $+1.0$ V. In addition, there is the possibility that the anion will complex with the central metal ion. This is especially true for the case of Fe(III), Fe(IV), and to some extent Fe(II).

Finally, there is the question of concentration of inert electrolyte. Potentials measured with concentrations of 1.0 M TBAP are not identical to those measured at 0.1 to 0.01 M. The shifts of potential may amount to several hundred millivolts and will vary as a function of the specific reactions examined.*

---

* Kadish, K. M., unpublished results.

D. SOLVENT AND COUNTERION EFFECTS ON REVERSIBLE HALF-WAVE POTENTIALS

Both the nature of the Fe(III) counterion and the solvent system have a strong effect on standard potentials for the reaction Fe(III)/Fe(II). A smaller effect is seen for the reactions Fe(II)/Fe(I), and almost no effect for reactions involving oxidation of Fe(III). This is discussed in detail in the following pages.

i. Fe(III) ⇌ Fe(II)

The solvent and counterion effect on the Fe(III)/Fe(II) reaction has been observed in a number of studies. In 1974, Lexa and coworkers [9] reported half-wave potentials for reduction of a number of TPPFeX complexes at a rotating platinum disk electrode. When $X = Cl^-$, Ac^-, or SCN^- in DMF the potential was observed to be -0.20, -0.20, or $+0.04$ V, respectively. A study by Wolberg [47] at about the same time reported potentials of -0.34, -0.21, and $+0.14$ V for reduction of the same complex in CH_2Cl_2, where $X = Cl^-$, Br^-, and ClO_4^-. A year earlier Kadish and Davis [48] had shown that TPPFeCl was reduced in DMSO at -0.09 V. All of these studies were carried out by cyclic voltammetry at a platinum working electrode.

Based on these initial studies, it was concluded that strongly coordinating solvents shifted the potentials in a positive (anodic) direction, while strongly coordinating anions shifted the potential in a negative (cathodic) direction. This effect of solvents on half-wave potentials was investigated in more detail by Constant and Davis in 1975 [49] and again by Kadish and Davis in 1976 [50]. In the latter study both solvent and substituent effects were investigated for the reactions of (p-X)TPPFeCl. The unsubstituted complex was reported to be reduced at -0.11, -0.11, and -0.15 V in DMSO, DMA, and DMF respectively. In poorly coordinating solvents such as CH_2Cl_2 or butyronitrile the half-wave potentials were shifted to -0.29 and -0.32 V. This anodic shift of $E_{1/2}$ with increased solvent coordinating ability was attributed to an increased stability of TPPFe versus TPPFeCl with coordinating solvents.

In 1981 the first complete systematic study of the interacting effects of solvent and axially coordinated monovalent anions on the electroreduction mechanisims and redox potentials of iron porphyrins was presented [46]. Five different anions were coordinated to $[TPPFe(III)]^+$, and their respective redox reactions were investigated in twelve different nonaqueous solvents. Potential shifts with changes in solvent were directly related to solvent donicity. In addition, potential shifts with changes in counterion were found, as expected, to be related to the degree of iron(III)-counterion interaction and were also solvent-dependent. Data taken from this study are reproduced in Table 5. Based on these data, a clear picture of the interdependence of solvent-counterion coordination with the Fe(III) center emerges. As seen from this table,

Table 5. Half-Wave Potential (V versus SCE) for the Fe(III)/Fe(II) Redox Couple of TPPFeX in Selected Solvents[a]

Solvent	Donor number	Dielectric constant	$E_{1/2}$ (X)				
			ClO_4^-	Br^-	Cl^-	N_3^-	F^-
EtCl$_2$	0.0	10.7	0.24	−0.19	−0.31	−0.38	−0.47
CH$_2$Cl$_2$	0.0	8.9	0.22	−0.21	−0.29	−0.42	−0.50
CH$_3$NO$_2$	2.7	35.9	0.10	b	b	b	b
φ-CN	11.9	25.2	0.20	−0.18	−0.34	−0.39	−0.57
CH$_3$CN	14.1	37.5	0.11	b	b	b	b
PrCN	16.6	20.3	0.13	−0.15	−0.27	−0.33	−0.45
(CH$_3$)$_2$CO	17.0	20.7	0.09	−0.16	−0.28	−0.34	−0.43
THF	20.0	7.6	0.17	−0.24	−0.34	−0.38	−0.47[a]
DMF	26.6	38.3	−0.05	−0.05	−0.18	−0.25	−0.40
DMA	27.8	37.8	−0.04	−0.05	−0.15	−0.24	−0.36
DMSO	29.8	46.4	−0.09	−0.09	−0.10	c	−0.09(−0.40)[d]
Py	33.1	12.0	0.15	0.17	0.16(−0.25)[d]	0.15(−0.28)[d]	0.16(−0.46)[d]

[a] Data taken from L. A. Bottomley and K. M. Kadish, Reference [46].
[b] Complex was insoluble in solvent system.
[c] Reduction consisted of two overlapping processes.
[d] Second reduction process.

reduction potentials for Fe(III) significantly shift as a function of counterion and solvent. In EtCl$_2$, reduction of TPPFeX becomes much more difficult (by up to 0.71 V) as the counterion varies from the weakly coordinating ClO$_4^-$ to the tightly bound F$^-$. This indicated preferential stabilization of the Fe(III) species over the Fe(II) species. The Fe(III) porphyrin-counterion binding strength increases in the order $ClO_4^- < Br^- < Cl^- < N_3^- < F^-$, and this is reflected in the half-wave potentials. Similar stabilizations of Fe(III) over Fe(II) by the counterion are observed in CH$_2$Cl$_2$ and φ-CN where the potential difference between the complexes were X = ClO$_4^-$ and X = F$^-$ is 0.72 and 0.77 V, respectively. In the weakly coordinating solvents DMA and DMF, this potential difference has decreased to 0.32 and 0.35 V, respectively. In DMSO and py, reduction of TPPFeX is essentially independent of the counterion. This invariance of $E_{1/2}$ with changes in X suggests identical reactants and products for all five of the complexes investigated and implies that DMSO or py solvent molecules have displaced the counterion from the Fe(III) center. Spectroscopic studies of the reduced complexes in py confirm complexation of Fe(II) by two py molecules. Studies of Fe(III) show that for TPPFeClO$_4$, the counterion is also replaced by two molecules [51]. For the remaining TPPFeX complexes in py, an equilibrium exists between TPPFe(py)$_2^+$ and TPPFeX or TPPFeX(py) [51].

The effect of solvent on the half-wave potentials for a number of simple metal ions and organometallic complexes has been discussed in the literature. In the case of iron porphyrins many of the effects may be attributed to solvent binding at the axial coordination position of the oxidized or reduced forms of the complex [49–52]. However, even in the absence of solvent coordination there are likely to be apparent shifts in $E_{1/2}$ for measurements made in different solvents versus the same standard calomel electrode. This is due to an inherent liquid-junction potential (l.j.p.) between the nonaqueous solution and the aqueous reference electrode. The magnitude of the l.j.p. may be substantial and will lead to an anodic shift of potential from the value observed in water. Because of this, corrections should be made for l.j.p. before attempting thermodynamic correlations between a variety of solvents. For example, in CH_2Cl_2, the l.j.p. has been measured as being up to 200 mV with respect to aqueous solutions. In DMSO the value is approximately 50 mV, and in DMF it is approximately zero [53].

A variety of methods are available in the literature for estimating liquid-junction potentials and obtaining corrected values of half-wave potentials [54–58]. The most often used, but not necessarily the most correct, is to measure the potential for the ferrocene/ferrocinium couple. Experimentally measured values of $E_{1/2}$ range between $+0.35$ and $+0.65$ V versus SCE, depending on the solvent system. In theory, this couple is independent of the solvent, so that any differences in observed potentials are due to the liquid junction. Values of $E_{1/2}$ of the porphyrins can then be corrected for the l.j.p. by adding or subtracting deviations in measured half-wave potentials of the Fc^+/Fc couple from its $E_{1/2}$ in H_2O. In practice, however, what is more often done is to report the potentials of the porphyrin reaction as $E_{1/2} - E_{1/2}(Fc^+/Fc)$ in each solvent.

Estimates of liquid-junction potential have occasionally been presented in the literature for a number of nonaqueous solvents. For a given solvent, these numbers will vary, often considerably, from laboratory to laboratory, depending on the method used in the calculation, the assumptions involved in the equations, and the individual experimental setup. These literature corrections should never be used to obtain thermodynamic correlations with metalloporphyrin redox potentials measured in a different laboratory under different experimental conditions. Values of the l.j.p. may vary substantially not only from laboratory to laboratory but even from day to day in a given laboratory, depending on the condition of the individual reference-electrode/solution junctions utilized in the experiment. Thus, although precisions of ± 1 mV may be obtained in the laboratory, the reporting of a measured $E_{1/2}$ to better than ± 10 mV makes little sense.

After all liquid-junction potentials have been corrected, it is possible to consider the effect of solvent complexation in determining iron porphyrin redox potentials. In the case of metalloporporphyrin reactions, solvent effects may be considered as similar to ligand effects in that shifts of potential will be either positive or negative, depending on the relative porphyrin-solvent interaction of each oxidation state. If the solvent molecule binds only to the oxidized form of the complex, a negative potential shift will be observed with respect to a noncoordinating solvent. In contrast, binding of the solvent to the reduced form, but not the oxidized form, will yield a positive shift in potential, with an increase in the binding ability of the solvent. Finally, if both the oxidized and reduced forms are complexed by solvent, the direction of potential shift will depend on the ratio of stability constants of each oxidation state.

Generally, for positively charged species, the higher oxidation state is stabilized with respect to the lower oxidation state and a negative shift of potential is observed. This was first shown to be true for the case of TPPCo oxidation to yield TPPCo(III) [59, 60]. In strongly coordinating solvents such as pyridine, a facile oxidation of Co(II) is observed at -0.21 V, while in solvents such as butyronitrile or benzonitrile the potential is shifted anodically to 0.395 and 0.486 V, respectively. The most positive potential for Co(II)/Co(III) is $+0.85$ V, which was obtained in CH_2Cl_2 [60].

The coordinating ability of aprotic solvents for cations decreases in the following order: pyridine > butyronitrile > benzonitrile > CH_2Cl_2. This follows the ease of Co(II) oxidation in each solvent. A more exact correlation, however, is available if $E_{1/2}$ is plotted versus the Gutmann donor number of the solvent [61].

In order to measure the donor ability of solvents Gutmann [61] measured their reaction enthalpies with $SbCl_5$ in 1,2-dichloroethane. The chemical reaction investigated was $S + SbCl_5 \rightleftharpoons SbCl_5 \cdot S$. The negative enthalpy of this reaction has been determined for a number of solvents and is termed the donor number (DN) of the solvent. Most electrochemical solvents are donor solvents. The numbers obtained range from 2.7 in nitromethane to 38.8 in hexamethylphosphoramide. Benzonitrile has a donor number of 11.9, while DMSO and pyridine are 29.8 and 33.1, respectively.

A number of polarity scales have been proposed in order to define the donor-acceptor properties of a given solvent and to rank solvents in order of increasing polarity. Often a combination of several terms is used. However, for the case of correlations with standard redox potentials the Gutmann donor number has proven to be the most successful as well as the most informative.

In a large number of cases plots of $E_{1/2}$ versus DN are linear, with a shift toward cathodic potentials as the donor number increases. This is most often true for positively charged species and indicates a stabilization of the higher oxidation state versus the lower state. Occasionally a linear plot is obtained with a slope in the opposite direction, indicating stabilization of the lower oxidation state.

In order to quantify the effect of solvent coordination for the reaction Fe(III)/Fe(II), a correlation was made between the half-wave potentials for reduction of TPPFeX and the donor number of the solvent. Such a plot is shown in Figure 5 for the reactions of $TPPFeClO_4$, TPPFeBr, $TPPFeN_3$, and TPPFeF. To eliminate differences in l.j.p. contributions arising between one solvent and another, the potentials were referenced against the measured value of the ferrocene/ferrocinium-ion couple in each solvent. The results obtained, and the differences between complexes where $X = ClO_4^-$ and where $X = Br^-$,

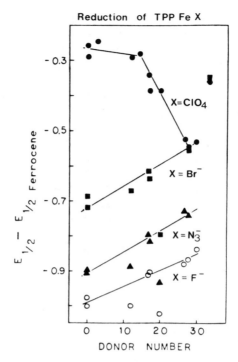

Figure 5 Half-wave potential for the reduction of TPPFeX plotted as a function of donor number. (Reprinted with permission of the American Chemical Society [46].)

N_3^-, or F^-, provide a lucid means for demonstrating the interacting effects of solvent and counterion on the electrode reactions of iron(III) and iron(II) porphyrins.

In solvents of donicity up to $\simeq 15$, the potential for reduction of TPPFeClO$_4$ remained essentially constant. This suggests that in these solvents ClO$_4^-$ is bound to the Fe(III) center and that there is only weak or no direct interaction between the solvent and the Fe. As the solvent donicity increased from 15, to Fe(III)/Fe(II) potential shifted negatively with increased DN. Pyridine is an exception to this trend, and with an observed potential of 0.15 V versus SCE is more positive than would be predicted.

In marked contrast to the effect of solvent on TPPFeClO$_4$ reduction, potentials of TPPFeX reduction, where X = Br$^-$, Cl$^-$, N$_3^-$, and F$^-$, are shifted in a *positive* potential direction by coordinating solvents. For X = Br$^-$, N$_3^-$, and F$^-$, the most negative potentials occurred in nonbonding solvents (THF excepted), while the most positive redox potential was observed in py. The magnitude and direction of the potential shift with solvent indicates a strong stabilization of Fe(II) relative to Fe(III). Stability-constant measurements of TPPFe and TPPFeCl bear this out. The value of $\log \beta_2$ for solvent binding to Fe(II) in CH_2Cl_2 is 7.45 [62] to 7.8 [52] for py, and 0.53 for DMF [62]. In contrast, TPPFeCl binds py only weakly, and DMF weakly or not at all [63].

The lack of fit by py to the trends observed for all of the complexes investigated may be an inherent property of the donor-atom–metal interaction, or it may be due to a change of spin state and/or coordination which is different in this solvent from all of the other solvents investigated. It is unlikely that the observed deviation is due to errors in the l.j.p. correction, since deviations from the predicted potential are also observed in similar plots for the reaction TPPCo(III) ⇌ TPPCo(II), but in this case in the opposite direction (see following sections). The deviation of potentials in THF for $X = Br^-$, Cl^-, N_3^-, and F^- but not ClO_4^- has also been discussed [46] and has been postulated as being possibly due to a difference of axial ligation between this solvent and all of the others. When a deviation of this type is observed from linear plots of $E_{1/2}$ versus DN, this may indicate a possible difference in chemistry for a given compound and suggest further experiments regarding its characterization.

ii. Fe(II) ⇌ Fe(I)

The insensitivity of the reaction Fe(II) ⇌ Fe(I) to changes in counterion on the starting iron(III) was observed by the groups of Lexa [9], Kadish [50–52], and Davis [49], as well as in the early work of Wolberg [47]. For the second reduction of TPPFeX in DMF where $X = Cl^-$, OAc^-, and SCN^-, Lexa [9] obtained an $E_{1/2}$ of -1.03, -1.03, and -1.02 V. Likewise measurements in CH_2Cl_2 by Wolberg [47] where $X = Cl^-$, Br^-, and ClO_4^- revealed an $E_{1/2}$ for Fe(II) reduction of -1.12, -1.16, and -1.18 V, respectively. The shift toward negative potential on going from DMF to CH_2Cl_2 is found for almost all reactions of metalloporphyrins where there is little or no complexation and is due to differences in both ion-pairing and liquid-junction potentials between the two solvents.

The reduction of TPPFe(II) also appeared from early studies to be relatively insensitive to changes in solvent, presumably because Fe(I) is noncomplexing. Reduction of either TPPFe(DMF) or $TPPFe(L)_2$ yielded the same iron(I) species, independent of the axial ligation on the iron(II) complex [10]. Identical spectra were obtained for reduction of iron(II) deuteroporphyrin complexes irrespective of the iron(II) axial ligation. These results, as well as later electrochemical data [9], lead to the postulate that Fe(I) complexes of porphyrins are not axially bound by nitrogeneous bases. What was not explained, however, is why two types of iron(I) spectra were generated in DMF [10]. These forms were identified as form A and form B and appeared to be in equilibrium. The optical spectra of both forms could be generated, in the case of $[TPPFe(I)]^-$, from either the monomeric starting product TPPFeX or the dimeric complex $(TPPFe)_2O$ [20]. Form A was identified as having a split Soret band at 457 and 362 nm, while form B also had a split Soret band with maxima at 420 and 390 nm. Form B is also formed by sodium amalgam reduction in THF [64].

That nitrogenous bases do not bind to Fe(I) appears to be clear-cut. Similar conclusions regarding solvent binding by Fe(I) cannot be made. In the case of THF as solvent, ESR spectra of Fe(I) were interpreted as having hyperfine structure arising from a bis THF adduct [64]. No hyperfine structure was displayed for Fe(I) in other solvents investigated.

Half-wave potentials for Fe(II) ⇌ Fe(I) in a variety of solvents are uninformative in elucidating the binding of iron(I) by solvent molecules. Reported values of $E_{1/2}$ ranged from -1.04 V in DMF to -1.14 V in DMSO for the same starting compound [50], but without corrections for l.j.p. exact thermodynamic differences as a function of solvent could not be ascertained. For this reason, a systematic study of the Fe(II)/Fe(I) reaction was undertaken [46]. Similarly to the studies of the reduction of Fe(III), Fe(II) reduction was investigated as a function of solvent and counterion. Five different counterions, F^-, N_3^-, Cl^-, Br^-, and ClO_4^-, were used in the complexes of TPPFeX and OEPFeX. For the former series, reductions were investigated in 10 different nonaqueous solvents.

The data are shown in Table 6, where values have not been corrected for l.j.p. As seen from this table, the sensitivity of $E_{1/2}$ to counterion depends upon the specific solvent. Previous studies on TPPFeCl have suggested that Cl^- dissociates from the Fe center after electron transfer to form TPPFe(II) [49]. Thus, electroreduction in noncoordinating media would involve a four-coordinated, intermediate-spin reactant, TPPFe(II), and a four-coordinated product, [TPPFe(I)]$^-$.

The similarity in observed potentials for the Fe(II)/Fe(I) couple in EtCl$_2$ when $X = ClO_4^-$, Br^-, and Cl^- suggests dissociation of the halide and

Table 6. Half-Wave Potentials of the Fe(II)/Fe(I) Redox Couple For TPPFeX in Selected Solvents[a]

Solvent	Donor number	Dielectric constant	$E_{1/2}$ (X)				
			ClO_4^-	Br^-	Cl^-	N_3^-	F^-
EtCl$_2$	0.0	10.7	-1.05	-1.06	-1.06	-1.11	-1.42
CH$_2$Cl$_2$	0.0	8.9	-1.06	-1.06	-1.07	-1.06	-1.50
φ-CN	11.9	25.2	-1.06	-1.06	-1.09	-1.12	-1.48
PrCN	16.6	20.3	-1.04	-1.03	-1.05	-1.06	-1.09
(CH$_3$)$_2$CO	17.0	20.7	-1.00	-0.99	-1.02	-1.06	-1.04
THF	20.0	7.6	-1.12	-1.15	-1.18	-1.13	-1.46
DMF	26.6	38.3	-1.03	-1.03	-1.03	-1.04	-1.02
DMA	27.8	37.8	-1.06	-1.05	-1.03	-1.08	-1.08
DMSO	29.8	46.4	-1.14	-1.14	-1.14	-1.14	$-1.15, -1.39$
Py	33.1	12.0	-1.50	-1.48	-1.51	-1.48	-1.50

[a] Data taken from L. A. Bottomley and K. M. Kadish, Reference [46].

identical coordination for the three complexes. However, the shift in potential to a more difficult reduction when $X = N_3^-$ or F^- implies the existence of a five-coordinated [TPPFe(II)X]$^-$ species. This was shown to be the case with F^-, and [TPPFeF]$^-$ was spectrally identified [46]. Similar arguments involving five-coordinated Fe(II) complexes can be made for electron transfers in CH_2Cl_2, ϕ-CN THF, and Me_2SO when $X = F^-$. However, in PrCN, $(CH_3)_2$, DMF, and DMA, the potentials of TPPFeF are essentially identical with TPPFeClO$_4$, indicating that the F^- is not coordinated to the Fe(II) electrode reactant.

In pyridine all TPPFeX complexes give essentially identical Fe(II)/Fe(I) potentials, which are shifted negatively with respect to the potentials observed in the other solvents. The donicities of py and DMSO are comparable (33.1 and 29.8, respectively), and one would expect similar stabilities of the TPPFe(py)$_2$ and TPPFe(DMSO)$_2$ complexes. The proposed existence of a negatively charged [TPPFeX]$^-$ form in DMSO in contrast with the absence of this species in py was attributed to the low dielectric constant of py (12.0 as compared to 46.4 for DMSO) [46].

iii. Oxidation of Fe(III)

Characterization of the formal oxidation state in oxidized iron(III) complexes has been an area of interest and controversy for 10 years. Suggestions have been made for π-radical cation formation based upon regularities of the half-wave potentials [65] and magnetic susceptibility measurements [66]. Recent NMR studies [67, 68] and Mössbauer studies [69, 70] also support the idea of π-radical formation for monomeric iron porphyrins.

In contrast, Felton and coworkers [5, 6] have identified the oxidation product as Fe(IV) for both TPPFeClO$_4$ and OEPFeClO$_4$ starting products. This assignment was based upon both electronic absorption spectra and magnetic-susceptibility measurements of the oxidized monomer. In this study a value of $S = 2$ at 200–300 K was assigned, as opposed to an earlier reported value of $S = 1$ obtained by NMR [66].

Oxidation of TPPFeCl by electrochemical methods gave an $E_{1/2} = 1.13$ V in CH_2Cl_2 [6, 47], while TPPFeBr and TPPFeClO$_4$ have half-wave potentials of 1.08 and 1.06 V, respectively [47]. The difference of 70 mV between TPPFeClO$_4$ and TPPFeCl cannot be considered significant, and it appeared that there was little influence of Fe(III) oxidation potential on counterion. Similar lack of solvent effects of Fe(III) oxidation potentials could not be ascertained, because the only values besides those reported in CH_2Cl_2 were obtained in butyronitrile. For the case of TPPFeCl oxidation, the potential was reported as 1.14 V [50].

In 1981 values were published for the oxidation of a large number of complexes of TPPFeX and OEPFeX in CH_2Cl_2 and, for the former complex, in six solvents. Measurements by Goff et al. [67] on 12 different TPPFeX

Table 7. Half-Wave Potentials (V versus SCE) for Oxidation of TPPFeX Complexes in Several Solvents[a]

Solvent	Donor number	Dielectric constant	$E_{1/2}(X)$				
			ClO_4^-	Br^-	Cl^-	N_3^-	F^-
$EtCl_2$	0.0	10.7	1.13	1.17	1.17	1.12	1.15
CH_2Cl_2	0.0	8.9	1.11	1.18	1.14	1.08	1.14
ϕ-CN	11.9	25.2	1.13	1.14	1.16	1.13	1.13
CH_3CN	14.1	37.5	1.12	b	b	b	b
PrCN	16.6	20.3	1.24	1.14	1.16	1.15	1.10
$(CH_3)_2CO$	17.0	20.7	1.17	1.17	1.17	1.16	—

[a] Data taken from L. A. Bottomley and K. M. Kadish, Reference [46].
[b] Complex was insoluble in solvent system.

complexes in CH_2Cl_2 were consistent with those obtained by Wolberg [47] for three TPPFeX complexes in the same solvent, and confirmed the invariance of $E_{1/2}$ with the starting counterion on Fe(III). Measurements by Kadish and Bottomley [46] were consistent with this interpretation in six different nonaqueous solvents. The data obtained are shown in Table 7. Because of limits in the anodic potential range, solvents were limited to those below a DN of 17 [$(CH_3)_2CO$]. As seen from this table, there is no appreciable shift in potential even though the coordinating ability of the solvent has increased. This behavior, as well as the invariance of $E_{1/2}$ with counterion [46, 47, 67], suggests that the electron-transfer site is ring-centered rather than metal-centered. Such an assignment is in agreement with the conclusions of Reed and coworkers [69, 70] as well as Goff et al. [67, 68]. Alternative explanations are that oxidation of TPPFe(III)X involves removal of an electron from a nonbonding orbital or that the process is metal-centered, but the relative stabilities of $[TPPFe(IV)X]^+$ ClO_4^- versus TPPFe(III)X are identical despite changes in X. Redox potentials alone are insufficient to differentiate among the above.

E. AXIAL LIGATION, REDOX POTENTIALS, AND CALCULATION OF FORMATION CONSTANTS

Two approaches have been used in investigating the relationship between redox potentials and axial ligation. The first, and the most simple, is to complex the porphyrin axially under known experimental solution conditions and to measure the redox potential under these conditions. Absolute values of $E_{1/2}$ may then be related to a given set of axial ligands or the iron spin state. The second approach is to complex the porphyrin axially (either the oxidized or reduced form) under a set of varying experimental conditions and, as a function of shifts of $E_{1/2}$ with these varying conditions (for example an increase in ligand concentration), determine the stoichiometry and formation constants for addition of the ligand to the porphyrin molecule. In the first case

Axial Ligation, Redox Potentials, and Calculation of Formation Constants 195

the interest is in determining a given $E_{1/2}$; in the second case half-wave potentials are used to determine the nature of the complex which exists under a specific set of experimental conditions and to evaluate the magnitude of the formation constants for the addition of one or two ligands to the porphyrin center.

Most studies of axial ligation have involved the reactions Fe(III) ⇌ Fe(II), some have involved the reactions Fe(II) ⇌ Fe(I), and a few have involved the oxidation of Fe(III). Perhaps the best understood of these reactions in terms of ligand and solvent effects on $E_{1/2}$ involve the reduction Fe(II) = Fe(I).

i. Iron(II) Complexation and the Reaction Fe(II) ⇌ Fe(I)

Iron(II) porphyrins may be four-, five-, or six-coordinated, depending on the solvent and the nature of the axial ligand. (See other chapters of this book). In contrast, iron(I) porphyrins will not readily bind axial ligands, and no Fe(I) porphyrin complexes with nitrogenous bases are known to exist. Thus, for reduction of Fe(II) complexes containing axially complexed nitrogenous bases the electrode reactions may be shown schematically as occurring by one of the three pathways presented in Figure 6. Since Fe(I) does not complex with nitrogenous bases, the potential, as predicted by the Nernst equation, will be directly related to the stability constant for either TPPFe(L) or TPPFe(L)$_2$ formation as well as to the concentration of free ligand in solution. The larger the stability constant, the larger the shift of $E_{1/2}$.

Numerous equations have been developed for the determination of stability constants using electrochemical methods [71]. The selection of a given equation will depend in large part on the specific electrochemical technique utilized as well as the magnitude of the overall stability constant and the difference

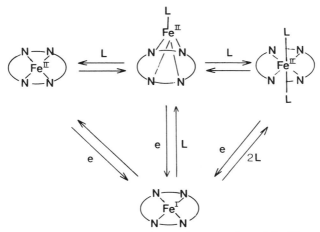

Figure 6 Possible electron-transfer pathways for reduction of Fe(II) porphyrins.

between the constants K_1 for addition of the first ligand and K_2 for addition of the second. For the specific case of Fe(II) porphyrins the equation which has most often been employed is

$$(E_{1/2})_c = (E_{1/2})_s - \frac{0.059}{n} \log K_{\text{Fe(II)}} - \frac{0.059}{n} \log (L)^p, \qquad (15)$$

where $(E_{1/2})_c$ is the half-wave potential for reduction of the complexed iron(II), $(E_{1/2})_s$ is the half-wave potential for reduction of the uncomplexed iron(II), $K_{\text{Fe(II)}}$ is the formation constant (either K_1 or β_2) of interest, (L) is the free-ligand concentration, and p is the number of ligands bound to Fe(II). For all iron-porphyrin complexes studied to date n has been 1, and p has been 0, 1, or 2. Equation (15) is perhaps better understood if one considers the reductions of PFe(L)$_p$ and PFe to yield the same [PFe(I)]$^-$ product, where P is a porphyrin. This is shown in scheme I below.

Scheme I: $(E_{1/2})_s$
$$\begin{array}{c} \text{PFe} \xrightleftharpoons{e} \\ {}_{pL}\updownarrow \qquad \xrightleftharpoons{e} [\text{PFe(I)}]^- \\ \text{PFe(L)}_p \xleftarrow{pL} \end{array}$$

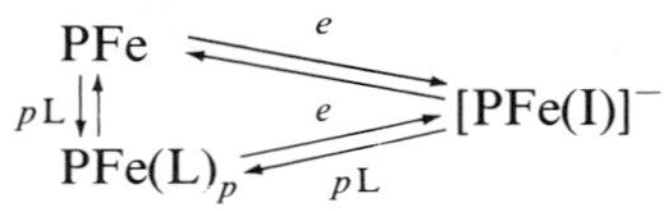

Calculation of the reversible half-wave potentials $(e_{1/2})_s$ and $(E_{1/2})_c$ will directly yield the formation constant for the ligand addition reaction to Fe(II). At 1.0 M ligand Equation (15) may be simplified to

$$\log K_{\text{Fe(II)}} = \frac{(E_{1/2})_c - (E_{1/2})_s}{-0.059 \text{ V}} \qquad \text{at } 25°C \qquad (16)$$

In practice one measures half-wave potentials for reduction of Fe(II) in the absence of complexing ligand, and then again in the presence of a series of different ligand concentrations. By appropriate selection of ligand stock solution, a titration may be performed and a number of half-wave potentials rapidly measured. An example of such a plot is given in Figure 7 for the reduction of OEPFeClO$_4$ in the presence of pyridine. From the slope of the plot, the value of p can be determined and $\log K_{\text{Fe(II)}}$ calculated after appropriate substitution into Equation (15). As seen in Figure 7, the curve has two segments, one in which the slope is zero, indicating $p = 0$ (no ligand binding) below $\log(\text{py}) = -2.0$, and another in which the slope is -120 mV, indicating $p = 2$ at all higher concentrations of pyridine. Further addition of pyridine will continue to shift the potential for reduction of the bis ligand adduct in a cathodic direction by $-(120 \text{ mV})/\log L$, although at high ligand concentrations additional shifts in $E_{1/2}$ will occur due to errors inherent in the liquid-junction potentials of the mixed solvent system. This does not usually occur until a ligand concentration of at least 1.0 M.

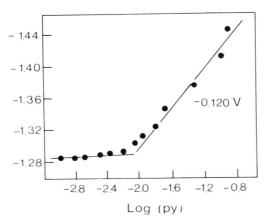

Figure 7 Plot of $E_{1/2}$ versus log(py) for the titration of OEPFeClO$_4$ with pyridine in CH$_2$Cl$_2$, 0.1 M TBAP. The electrode reaction shown is Fe(II) ⇌ Fe(I).

Examples of shifts of $E_{1/2}$ due to complexation of Fe(II) have been given by Lexa et al. [9] for imidazole and histidine complexation, by Constant and Davis for solvent complexation (including pyridine) [49], and by Kadish et al. for substituted pyridines [52], halide ions [46], solvent molecules [46], and diatomic molecules [72].

Complexation of Fe(II) by either imidazole to form bis ligand adducts, or sterically hindered imidazole to form mono ligand adducts, yields stability constants in the range of 10^4 and 10^7 depending on the solvent. Since $n = 1$, this corresponds to a 240–420-mV cathodic shift of potential [at (L) = 1 M] from that observed for the uncomplexed molecule [cathodic shifts are approximately (60 mV)/log K]. Thus, for example, TPPFe is reduced at -1.03 V in DMF [9], and TPPFe(Im)$_2$ is reduced at -1.20 V in the same solvent when the imidazole concentration is 6.5×10^{-3} M. In a similar manner, TPPFe has a log β_2 for pyridine complexation of 7.45 in benzene [62] and 7.8 in CH$_2$Cl$_2$ [52]. In pyridine $E_{1/2} = -1.50$ V (see Table 6).

The relationship between the ligand-binding strength of the metalloporphyrin and $E_{1/2}$ was best shown in a paper by Kadish and Bottomley [52], where $E_{1/2}$ was plotted versus the pK_a of the axial ligand. Eleven substituted pyridines were investigated between pK_a 0.67 and 9.71. For all of the ligands in which a wave was observed, $E_{1/2}$ was directly proportional to the pK_a of the ligand. The most positive reduction potential occurred for TPPFe(L)$_2$ where L = 3,5-dichloropyridine ($E_{1/2} = -1.38$ V, log $\beta_2 = 5.5$), while the most negative occurred where L = 4-picoline ($E_{1/2} = -1.55$ V, log $\beta_2 = 8.3$). 4-(dimethylamino)pyridine was estimated to have log $\beta_2 = 9.7$, and for this complex $E_{1/2}$ has shifted so that there is an overlap of the Fe(II)/Fe(I) reaction with the ligand-centered reaction.

Stability constants and half-wave potentials for reduction of intermediate spin iron(II) have also been measured for complexes containing a number of oxygen donors, sterically hindered nitrogenous bases, and in several cases halide ions in nonbonding solvents [74]. In addition the electrode-reaction and electronic-absorption spectra of Fe(II) and Fe(I) have been monitored during titrations of TPPFeX with several imidazoles and pyridines [75]. Values of K_f determined electrochemically for TPPFeL formation vary between 10 and 200 for sterically hindered pyridines and between 10^4 and 10^5 for sterically hindered imidazoles. For DMF, EtOH, and MeOH complexation, literature values are available from spectrophometric data [62]. In these cases, good agreement is obtained between the electrochemical and spectroscopic techniques. In order to accomplish an electrochemical determination of log K_1 for these weakly binding solvent molecules, the appropriate ligand (solvent) was titrated into CH_2Cl_2 solutions containing the iron porphyrin, and the potential recorded. Observed shifts of $E_{1/2}$ were usually less than 60 mV and gave values of log K_1 or log β_2 less than 10.

In one instance [46], the binding of a halide to Fe(II) could be measured. Formation of TPPFeF$^-$ was observed electrochemically and the spectra recorded during titration of TPPFe with F$^-$. Isosbestic points were obtained, and a well-defined shift of potential from -1.06 V to > -1.50 V was obtained as a function of (F$^-$). [TPPFeF]$^-$ may also be isolated and characterized by physical-chemical measurements [76].

The delicate balance between redox potentials and stability constants for the oxidized and reduced forms of a complex are nicely illustrated in the reduction of TPPFe(II) which is complexed by a diatomic molecule such as NO or CO. Complexes of Fe(II) with NO which have been characterized include the low-spin mono adduct, PFe(NO) [77], the low-spin bis adduct PFe(NO)$_2$ [77], and the mixed nitrogenous base NO adduct PFe(NO)(L) [78–80]. Complexes of Fe(II) with CO include the mono and bis CO complexes as well as the mixed CO-L species [81–83]. All of these complexes may be electrochemically reduced and their reduction products characterized by electrochemical and spectroscopic methods [84]. Potentials for reduction of several NO and CO complexes are listed in Table 8.

In either CH_2Cl_2 or pyridine the nitrosyl complex of TPPFe and OEPFe could be reversibly reduced [72]. In a nonbonding solvent the electrode reactions are

$$PFe(NO) + e \rightleftharpoons [PFe(NO)]^-,$$

while in a bonding solvent the reaction becomes

$$PFe(NO)(py) + e \rightleftharpoons [PFe(NO)]^- + py.$$

Table 8. Half-Wave Potentials for Oxidation and Reduction of Several Iron-Porphyrin Complexes with Diatomic Molecules, in CH_2Cl_2

Complex	Diatomic	Other ligand	$E_{1/2}$(ox)	Ref.	$E_{1/2}$(red)	Ref.
TPPFe	NO	—	0.74	[72, 138]	−0.93	[72]
	CO	—	0.60[a]	[84]	−1.22	[84]
	NO	py	0.54[a]	[72, 138]	−0.98	[72]
	CO	py	0.50[a]	[104]	—	—
	CO	Im	0.38[a]	[137]	—	—
OEPFe	NO	—	0.60	[72, 103]	−1.10	[72]
	CS	—	0.73	[103]	—	—
	NO	py	0.57[a]	[103]	—	—
	CO	py	0.43[a]	[103]	—	—
	CS	py	0.58	[103]	—	—
	CS	1-MeIm	0.52	[103]	—	—
	CS	pip	0.58[a]	[103]	—	—
EtioFe	CO	Im	0.32[a]	[102]	—	—

[a] Irreversible reduction. The value presented is $E_{p,a}$.

Evidence for coordination of NO to the reduced species was based on both the invariance of $E_{1/2}$ as a function of NO pressure and the electronic absorption spectra [72]. It is of interest to note that the observed potentials $E_{1/2} = -0.93$ V in CH_2Cl_2 and -0.98 V in pyridine indicate a greater stability of the reduced NO complex than of the starting material. This is in agreement with theoretical predictions [72]. No evidence exists for formation of an Fe(I) species in this reduction, and the reduced product might be formulated as a direct reduction of bound NO ligand. The reduction products of PFe(NO) were not isolated, even though this was a well-defined reversible one-electron reduction in the cyclic-voltammetry experiment. Controlled-potential coulometry in CH_2Cl_2 at -1.2 V indicated a catalytic reaction involving [PFeNO]⁻, which presumably reacted with solvent to regenerate the starring material.

The two complexes TPPFe(CO) and TPPFe(CO)₂ may also be reversibly reduced in nonaqueous solvents. In this case however, the potential is shifted in a cathodic direction, indicating greater stability of the Fe(II) species with CO. It is not clear at this time if a reduced [TPPFe(CO)]⁻ species is formed. There is no difference between potentials for anion radical formation under a CO atmosphere and in the absence of CO and this, however, might suggest against a [TPPFe(CO)]⁻ species.

ii. Iron(III) Complexation and the Reaction Fe(III) ⇌ Fe(II)

The most studied reaction of iron porphyrins has been that for Fe(III)/Fe(II), and a number of papers have been published showing how this potential will change as a function of various experimental conditions. The main thrust of many of these studies has been to correlate changes in $E_{1/2}$ with changes in electron-transfer mechanisms and pathways, and at the same time to elucidate structure-function relationships. For the latter, attempts have been made to correlate changes in $E_{1/2}$ with metal spin state, axial-ligand coordination, solvent system, counterion of Fe(III), and basicity of the porphyrin ring.

Iron-porphyrin complexes of Fe(III) and Fe(II) may be four-, five-, or six-coordinated in solution, and will contain either high-, intermediate-, or low-spin Fe depending upon the type of axial ligand. All six-coordinated iron-porphyrin complexes with nitrogenous bases have been identified as being low-spin [85], whereas all six-coordinated complexes with oxygen-donor ligands are high-spin [86]. The intermediate-spin state has been assigned to TPPFe(II) [87–89] when the porphyrin is dissolved in nonbonding aprotic media. This implies a vacant fifth and sixth axial position. For Fe(III) with weakly coordinating anions (e.g., TPPFeX where X = ClO_4^-, BF_4^-, or PF_6^-), an admixture of $S = \frac{3}{2}, \frac{5}{2}$ has been assigned to the system [86, 90]. Low-spin pentacoordinated complexes are rarely observed. In contrast, high-spin Fe(III) complexes are formed when the associated counterion is either a halide or a strong-field anion (e.g. N_3^-). High-spin porphyrin complexes of Fe(II) are also formed when a sterically hindered nitrogenous base occupies the fifth coordination site [91].

Thus, the exact nature of the axial coordination will produce a given Fe(III) or Fe(II) complex in a given spin state which may have completely different redox properties than another very similar complex. In addition, slight changes in coordination geometry may lead to large changes in the metal–porphyrin-plane distance and correspondingly large changes in the redox characteristics. An attempt to determine the exact effect of axial ligation on half-wave potentials is thus not a trivial matter, although potentials under a given experimental set of conditions can be reported.

When considering all possible spin states of each reactant and product, nine paths of electron transfer are possible between iron(III) and iron(II). If several different complexes exist in the same spin state, the number increases. In fact many different types of electron-transfer reactions are possible between a simple iron(III) and an iron(II) porphyrin in solution. This is shown in Figure 8, which illustrates several pathways that may occur for the reduction of monomeric PFeX, where X is a counterion, in the presence of a simple nitrogenous base or complexing solvent. Almost all of these reactions have been identified under a given set of experimental conditions.

Similar schemes may be written for the reduction of monomeric iron porphyrins containing diatomic molecules in nonaqueous media (shown in Figure 9 for NO), monomeric iron porphyrins containing mixed sulfur-nitrogenous base ligands in nonaqueous media, and dimeric iron porphyrins of various oxidation states.

A consideration of Figures 8 and 9 easily convinces one that the possible reaction schemes for Fe(III) and Fe(II) are numerous. It should be pointed out, however, that the schemes shown in Figures 8 and 9 correspond only to the

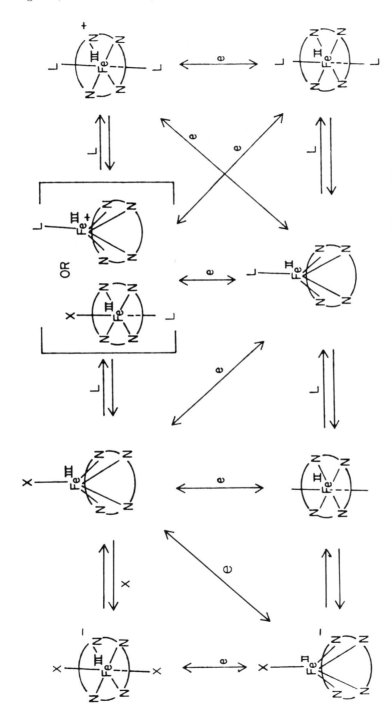

Figure 8 Schematic showing possible reduction paths from Fe(III) to Fe(II).

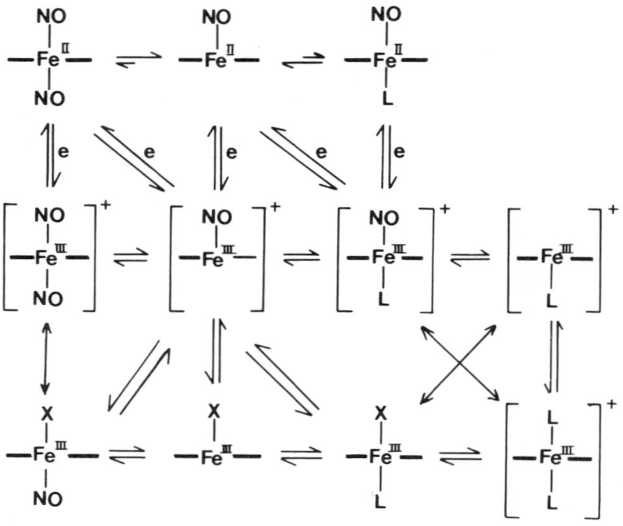

Figure 9 Schematic showing possible oxidation paths from Fe(III) to Fe(II) where Fe(II) is complexed by at least one NO molecule.

actual initial reactants and final products in the electrode reaction Fe(III) ⇌ Fe(II). These are the species which determine the reversible half-wave potentials according to the Nernst equation. As has been shown in many cases, a simple electron-transfer pathway is not always followed, and electron transfers along a quite different pathway than those shown in Figures 8 and 9 may be responsible for determining the observed redox potentials. However, when considering the initial reactant and the final product, several conclusions may be reached in regard to the effect of axial ligation on half wave potentials.

Generally, π-donors stabilize the iron(III) state with respect to the iron(II) state. However, when the iron(II) spin state is converted to low spin and the iron(III) remains high-spin, Fe(II) is stabilized with respect to Fe(III). When both spin states are low-spin, Fe(III) is again stabilized relative to Fe(II). Strong π-donors will convert both Fe(III) and Fe(II) to low spin and stabilize both spin states, with a slight favoring of Fe(II).

In all cases the half-wave potential for the reaction Fe(III) ⇌ Fe(II) will depend on the relative stabilization of Fe(III) versus Fe(II). For example, if both Fe(III) and Fe(II) are bis ligated, the following relationships may be observed:

$\beta_2(\text{PFe}(L)_2^+) > \beta_2(\text{PFe}(L)_2)$, negative shift of $E_{1/2}$;

$\beta_2(\text{PFe}(L)_2^+) < \beta_2(\text{PFe}(L)_2)$, positive shift of $E_{1/2}$;

$\beta_2(\text{PFe}(L)_2^+) \simeq \beta_2(\text{PFe}(L)_2)$, no shift of $E_{1/2}$.

Axial Ligation, Redox Potentials, and Calculation of Formation Constants

The exact relationships between half-wave potentials and stability constants is given by equation (17) (where $K = \beta_2$)

$$\log \frac{K_{Fe(III)}}{K_{Fe(II)}} = \frac{(E_{1/2})_c - (E_{1/2})_s}{-0.059 \text{ V}} \quad (17)$$

for the specific case of bis ligand formation by Fe(III) and Fe(II), and an example of how $E_{1/2}$ shifts with the ratio of stability constants is given in Figure 10.

Electrochemical calculations of stability constants for Fe(III) follow the same general pattern as described earlier. In this case calculations are more complicated in that both iron(III) and iron(II) will bind axial ligands, and the relevant electrochemical equation (where $K = \beta_1$ or β_2) becomes

$$(E_{1/2})_c = (E_{1/2})_s - 0.059 \log \frac{K_{Fe(III)}}{K_{Fe(II)}} - 0.059 \log(L)^{p-q} \quad (18)$$

at 25°C. In this case p is equal to the number of ligands axially bound to Fe(III) and q the number bound to Fe(II). Again, half-wave potentials are measured for the Fe(III) $\rightleftharpoons$ Fe(II) reaction where no ligands are bound and for reduction in the presence of complexing ligand. An example of such a titration is shown in Figure 11 for reduction of TPPFeCl in the presence of the sterically hindered ligand 1,2-dimethylimidazole. The electrode reaction at low

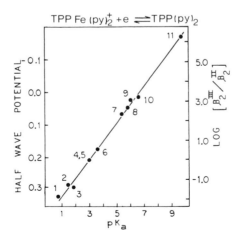

Figure 10 Plot of $E_{1/2}$ and $\log(\beta_2^{III}/\beta_2^{II})$ versus pK_a for the reduction of TPPFe(L)$_2^+$ in CH$_2$Cl$_2$ where L = substituted pyridine. Data taken from Reference [46].

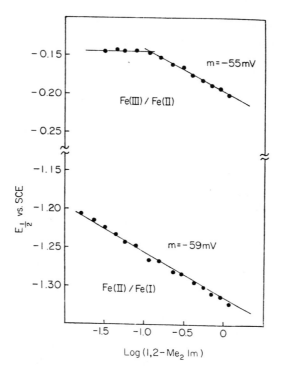

Figure 11 Plot showing $E_{1/2}$ for TPPFeCl reduction during titration with the sterically hindered ligand 1,2-dimethylimidazole.

ligand concentration is

$$\text{TPPFe(L)}^+ \underset{}{\overset{e}{\rightleftharpoons}} \text{TPPFe(L)}. \tag{19}$$

At higher ligand concentrations the slope is -55 mV, and the reaction in this region becomes

$$\text{TPPFe(L)}_2^+ \underset{}{\overset{e}{\rightleftharpoons}} \text{TPPFe(L)} + \text{L}. \tag{20}$$

The stoichiometry of complexation by Fe(II) and reduction to Fe(I) is consistent with this mechanism and is also shown in Figure 11.

Similarly to Equation (16), at 1.0 M ligand concentration the difference of potential between the complexed and uncomplexed forms will yield the ratio of stability constants as shown in Equation (17). This same equation is valid for solutions containing 1.0 M ligand as well as for the case when the slope of $E_{1/2}$ versus log(L) is zero, that is, when $p = q = 0$.

The importance of Equation (17) is that the actual potential observed will depend on the relative stability constants of the oxidized and reduced forms of the complex. This fact is often overlooked in trying to obtain a specific redox potential for a specific set of axial ligands on either Fe(II) or Fe(III). For example, in CH_2Cl_2, $K_{Fe(II)} > K_{Fe(III)}$ and a positive shift of potential is observed for complexation of TPPFeCl by pyridine. In contrast, when starting with TPPFeClO$_4$, $K_{Fe(II)} < K_{Fe(III)}$ and a negative shift of potential is obtained for pyridine complexation. Finally, when the ligand is 3-bromopyridine and the porphyrin TPPFeClO$_4$, $K_{Fe(III)} \simeq K_{Fe(II)}$ and no shift at all will be observed [52].

Because Equations (17) and (18) provide only a ratio of stability constants, it is necessary to utilize a separate measurement to explicitly determine the values for Fe(III) and Fe(II). This may be done by one of two methods, both of which involve an independent measurement of one of the two stability constants. If the reduction of Fe(II) to Fe(I) may be observed, $K_{Fe(II)}$ can be calculated in the manner explained and inserted into the ratio. If this reduction cannot be obtained, a spectral determination (preferably under the same experimental conditions) may be undertaken on either Fe(III) or Fe(II). Since Fe(III) is the air-stable form of the complex, it is usually easier to use.

The first use of modern electrochemical techniques to measure the relative stability constants and half-wave potentials for Fe(III)/Fe(II) containing axially complexed nitrogeneous bases in nonaqueous media was published by Lexa and coworkers [9] in 1974. In this study TPPFeX and the iron(III) complex of deuteroporphyrin IX dimethylester were utilized as the starting porphyrin. Bis ligand adducts were formed with either simple imidazole addition or by preparation of a heme peptide containing bis 1-histidine. The results for these two complexes were similar and were dependent on the counterion, as might be expected. When OAc$^-$ or Cl$^-$ were the counterions, complexation by imidazole produced bis adducts of Fe(III) and Fe(II) and a concommitant anodic shift of the standard potential from that of PFeX by approximately 80 to 120 mV. When X = SCN$^-$, however, the potential was shifted cathodically by approximately 100 mV from the starting potential. This shift, which was not explained at the time, is now better understood in terms of the counterion effect on PFeX potentials. Potentials for reduction of PFe(L)$_2$$^+$ X$^-$ will be fairly constant for a given value of X, while those for PFeX can vary by up to 720 mV depending on the solvent [46]. The magnitude of the potential shift for PFeCl and PFeOAc which was observed by Lexa indicates a greater stability for Fe(II) than for Fe(III), while that for PFeSCN indicates a greater stability for Fe(III) than for Fe (II).

This difference may be best understood by considering the measured values for imidazole addition to Fe(III) as a function of halide complexation. For example, in DMF the halide ion is only weakly bound to Fe(III), and in this solvent, $\log \beta_2$ for imidazole addition has been measured as between 7×10^6 and 2×10^7 and is virtually independent of the Fe(III) counterion [9]. In contrast, in nonbonding solvents the addition of imidazole to Fe(III) most clearly involves halide displacement:

$$PFeX + 2L \rightleftharpoons PFe(L)_2^+ X^-, \qquad (21)$$

and $\log \beta_2$ will vary as a function of X. In chloroform, values of $\log \beta_2$ for the reaction (21) where L = N-MeIm have been measured as 1.5×10^3 for X = Cl$^-$ and 2.5×10^7 for X = Br$^-$ [92]. Virtually identical final spectra of the bis ligand adduct are obtained, independent of the counterion. This is agreement with the conclusions derived from the invariance in redox potential for reduction of TPPFe(L)$_2^+$X$^-$ to yield TPPFe(L)$_2$ [52]. In this study identical values of $E_{1/2}$ were obtained for X = Cl$^-$ and X = ClO$_4^-$.

Recently Kadish and Bottomley [84] measured stability constants for the addition of imidazole and N-MeIm to TPPFeClO$_4$ in CH$_2$Cl$_2$ using electrochemical methods. The stability constants for both ligands were greater than 10^{14}. These are the highest ever measured for an iron metalloporphyrin system and are about 10^6 to 10^8 times as high as for similar complexes of TPPFeCl and OEPFeCl.

Half-wave potentials for the reduction of a series of bis ligated complexes with substituted pyridines and imidazoles were measured by Constant and Davis [49, 93] in DMSO and DMA. As would be expected, cathodic shifts of $E_{1/2}$ for Fe(III)/Fe(II) were obtained with increased pK$_a$ of the free ligand. Plots of $E_{1/2}$ versus pK$_a$ were not linear [49], due presumably to slow electron-transfer kinetics and to the fact that pyridines, imidazoles, and sterically hindered imidazoles were all plotted together on the same graph. Similar cathodic shifts of $E_{1/2}$ were also observed for the reduction of TPPFe(L)$_2^+$X in CH$_2$Cl$_2$ where X = ClO$_4^-$ or Cl$^-$, and L was a non-sterically-hindered substituted pyridine [52]. In this study formation constants for both Fe(III) and Fe(II) were calculated. A summary of the stability constants and half-wave potentials is given in Table 9.

Values of $\log \beta_2$ ranged between 5.5 and 9.7 for TPPFe(L)$_2$ formation and between 3.4 and 16.3 for TPPFe(L)$_2^+$Cl$^-$ formation. As would be predicted, half-wave potentials for reduction of Fe(III) to Fe(II) shifted cathodically by 60 mV for each tenfold increase in the ratio of $\log(\beta_2^{III}/\beta_2^{II})$. For all eleven nonsterically-hindered ligands a linear relationship was observed between $E_{1/2}$ and pK$_a$ of the ligand. The 480-mV difference in half-wave potentials for the Fe(III)/Fe(II) reduction can be associated with a 10^8 increase in higher-oxidation-state stability between the most difficult-to-reduce complex (containing 4-(N,N-dimethylaminopyridine) and that containing 3,5-dichloropyridine. Potential shifts similar in direction, but of smaller magnitude, also occurred for electrode reactions involving iron(II) and iron(I), and have been discussed earlier. For the case of Fe(III)/Fe(II) it is interesting to note that $\log(\beta_2^{III}/\beta_2^{II}) = -2.1$ when the axial ligand is 3,5-dichloropyridine, 0.0 for 3-bromopyridine, and $+6.5$ for 4-(dimethylamino)pyridine. The plot of $E_{1/2}$ versus pK$_a$ which has been obtained is shown in Figure 10.

Table 9. Half-Wave Potentials and Formation Constants for Bis Pyridine Complexes of TPPFe(III) and TPPFe(II)[a]

Ligand L	pK_a	$E_{1/2}$ (V versus SCE)		$\log \beta_2^{III}$ [b]	$\log \beta_2^{II}$ [c]
		Fe(III)/Fe(II)	Fe(II)/Fe(I)		
3,5-dichloropyridine	0.67	0.31	−1.38	3.4	5.5
3-cyanopyridine	1.40	0.29	−1.43	4.7	6.2
4-cyanopyridine	1.86	0.30	−1.32	2.6	4.2
3-chloropyridine	2.81	0.21	−1.46	6.7	6.7
3-bromopyridine	2.84	0.21	−1.45	6.8	6.8
4-acetylpyridine	3.51	0.18	N.O.[d]	7.6	7.0
pyridine	5.28	0.06	−1.52	10.2	7.8, 7.45[e]
3-picoline	5.79	0.05	−1.52	10.5	7.8
4-picoline	5.98	0.02	−1.55	11.4	8.3
3,4-lutidine	6.46	0.01	N.O.[d]	11.7	8.3
4-(dimethylamino)pyridine	9.71	−0.17	N.O.[d]	16.3	9.7

[a] Data taken from K. M. Kadish and L. A. Bottomley, Reference [52].
[b] Reaction: TPPFeClO$_4$ + 2L ⇌ TPPFe(L)$_2^+$ ClO$_4^-$.
[c] Reaction TPPFe + 2L ⇌ TPPFe(L)$_2$.
[d] N.O. = not observable.
[e] Taken from Reference [62].

iii. Mixed Axial Ligation of Fe(III)

Scattered reports of reduction potentials for synthetic iron(III) porphyrins with axial N and S donors have appeared in the literature [94–96]. In all cases, the ferrous form of the mixed ligand complex is stabilized over the ferric form in comparison with the similar bis nitrogenous base complexes. Using cyclic voltammetry in 30% dioxane-water mixtures, Wilson [94] measured the potentials for a series of synthetic meso heme derivatives containing either methionine, histidine, or imidazole as axial ligands. Preferential stabilization of the Fe(II) oxidation state was obtained for the mono methionine, bis methionine, and histidine-methionine derivative, in comparison with the similar bis histidine complex. The potential difference between the bis histidine mesoheme and the histidine-methionine mesoheme was 144 mV, the latter complex being oxidized at −0.074 V versus NHE. A similar 150 mV anodic shift of potentials was observed by Harbury [95] for an aqueous octapeptide system. Likewise, Reed and coworkers [96], using cyclic voltammetry in a low-dielectric-constant medium, obtained an anodic 167 mV shift of potentials when imidazole was replaced by a thioether molecule as axial ligand. In this instance the synthetic Fe(II) porphyrin investigated was a "tailed porphyrin" derived from meso (o-aminophenyl) triphenylporphyrin.

The potential difference between bis histidine cytochrome c_3 (−0.200 V versus SHE) [97] and cytochrome c (+0.260 V) [98] containing histidine and methionine as axial ligands is 460 mV. Thus, based on the results with thioether and covalently attached imidazole as axial ligands, it was concluded that replacement of histidine by methionine results in a 160-mV anodic shift of potential, with the remaining 300 mV ascribed to the protein environment about the heme. This is an interesting conclusion but has not been investigated in more detail. In fact, very little data exist for iron-porphyrin complexes containing mixed ligand systems. Because of the greater affinity of nitrogen than sulfur for the metal center [99], mixed ligand complexes of Fe require a tailed-porphyrin approach, wherein the nitrogen base is appended to the ring. This method has proven successful in the past [99–101] and will most likely find much utilization in future studies.

iv. Diatomic-Molecule Adducts

While the redox behavior of five-coordinated iron-porphyrin halides or pseudohalides is now well characterized, one area that has received only scant attention has been the electrochemical oxidations of metalloporphyrins with diatomic ligands such as CO, NO, O_2, or CS. These adducts are capable of dramatic shifts in the energies of the metal orbitals because of their π-accepting properties.

Iron(II) complexes with nitrosyl, thiocarbonyl, carbonyl, and dioxygen complexes have been well characterized in the literature. The first two complexes are stable in both the solid and the solution form and, in contrast to most other types of synthetic iron porphyrins, are known to stabilize the iron(II) state as the air-stable state. Thus electrode reactions involving complexes of NO and CS will involve oxidations of Fe(II) $\rightleftharpoons$ Fe(III). In contrast, the CO and O_2 adducts with Fe(II) depend upon the partial pressure above the solution, and can be characterized as either Fe(III)/Fe(II) or Fe(II)/Fe(III).

The irreversible oxidation of EtioFe(CO)(Im) was initially reported by Brown et al. [102] at $E_p = +0.32$ V in CH_2Cl_2. A similar irreversible oxidation was also observed at $E_p = +0.43$ V by Buchler [103] for OEPFe(CO)(py) as well as for OEPFe(NO)(py) at $E_p = +0.57$ V. In contrast, OEPFe(NO) is reversibly oxidized in CH_2Cl_2 [103]. Electrochemical oxidations of OEPFe^{II}(CS) and OEPFe^{II}(CS)(py) are also reversible, with the former complex shifted to potentials slightly positive from those of the NO and CO complexes [103]. Generally, axial coordination of one of the diatomic π-acceptor ligands shifts the Fe(II)/Fe(III) oxidation potential in an anodic direction by 500 to 1270 mV from that observed for TPPFe^{II} or TPPFe^{II}(L)$_2$ [104]. Similar anodic shifts, through smaller in magnitude, have been observed upon complexation of iron protoporphyrin IX with CO in aqueous media [105]. Recently, Kadish et al. [72, 84] reported the first detailed investigation of the electrode reactions of Fe(II) axially complexed by CO and NO. The oxidations of the TPPFe(NO) and TPPFe(NO)(L) were complex and have been described in Figure 9. A list of oxidation potentials is given in Table 8.

In the absence of competing ligands the paramagnetic six-coordinated [TPPFe(NO)$_2$]$^+$ and [OEPFe(NO)$_2$]$^+$ could be produced in solution by controlled-potential electrolysis. The product was isolated and characterized by electronic-adsorption spectroscopy, i.r. spectroscopy, and magnetic-susceptibility measurements [72]. TPPFe(CO), like the isolelectronic [TPPFe(NO)]$^+$, is quite labile and has not been isolated in the solid state. Reversible oxidation of TPPFe(CO) at potentials more positive than $+0.6$ V may be obtained at low temperatures [106]. At room temperature, however, the electron-transfer mechanism involves a rapid dissociation of CO from [TPPFe(CO)]$^+$ after the electron transfer to yield [TPPFe]$^+$.

v. Oxidation of Fe(III)

Ligand binding to oxidized Fe(III) by nitrogenous bases has been studied only slightly. Unpublished NMR results by the laboratory of Goff and electrochemical results by the laboratory of Kadish indicate that imidazole and pyridine will bind to the oxidized complex at low ligand concentrations, but that at high concentrations an internal electron transfer is observed and the oxidized species is reduced to the neutral complex TPPFe(L)$_2{}^+$. Detailed electrochemical experiments in the presence of nitrogenous bases are difficult to perform in that the electrode reaction is just at the anodic limit of most solvent systems containing nitrogenous bases.

F. PORPHYRIN STRUCTURE AND HALF-WAVE POTENTIALS

In order to use electrochemistry as a tool in the study of structural problems, correlations must be found between electrochemical reactivity and structure. In many cases one is able to express correlations between reactivity and structure quantitatively, while in others only qualitative descriptions may be given.

A number of quantitative treatments in porphyrin chemistry have involved the use of linear free-energy relationships. This has been discussed in some detail by Adler in the introduction to a conference on porphyrins organized by the New York Academy of Sciences in 1973 [107]. Dozens of correlations involving structure-reactivity relationships of metalloporphyrins were presented in 1973, and numerous others have appeared in the years since then. Included in these studies are many investigations of substituent effects on the physical and chemical properties of metalloporphyrins.

The effect of substituents in determining iron-porphyrin half-wave potentials has not been investigated in detail for a large number of complexes, and even in these cases the data are not always clear and depend on the specific axial ligand(s). It is well known that changes in ring substituents of an iron porphyrin will result in a change of electron density on the four nitrogens of the metalloporphyrin and will hence influence the physical properties and chemical reactions of a given porphyrin. At the same time, however, as a given set of axial ligands is varied, the influence of the π-ring system on the metal-centered reaction is also expected to vary. "Cis-trans" effects on the physical properties and chemical reactions of porphyrins have been extensively discussed by Buchler et al. [104] and Caughey et al. [108, 109]. There is an inverse relationship between porphyrin-metal and axial-ligand–metal interactions: the stronger the one, the weaker the other. An example of this is given by the Fe(III)/Fe(II) reactions of TPPFeX and OEPFeX.

As already seen in Table 5, half-wave potentials for Fe(III) ⇌ Fe(II) may vary by up to 720 mV for a TPPFeX complex, depending on the nature of the counterion, X, or the specific solvent system. Some data are reproduced in Table 10, which compares $E_{1/2}$ for reduction of OEPFeX and TPPFeX in three solvents. In the first, CH_2Cl_2, there is no solvent complexation of Fe(III) or Fe(II), while in the second, DMF, either Fe(III) or Fe(II) may be complexed by one or more solvent molecules. In the last solvent, pyridine, Fe(II) forms bis adducts completely, while Fe(III) complexes are coordinated by 15–100% [51]. As seen from this table, the shift of potentials observed in CH_2Cl_2 is substantial. For both OEPFeX and TPPFeX, a greater than 700-mV cathodic-potential shift is observed on going from $X = ClO_4^-$ to $X = F^-$. The Fe(III) porphyrin-counterion binding strength increases in the order $ClO_4^- < Br^- < Cl^- < N_3^- < F^-$, and this is reflected in the half-wave potentials. A parallel trend is observed between the OEPFeX and TPPFeX complexes. The absolute difference of potentials between TPPFeX and OEPFeX for a given X is a constant 0.12 ± 0.01 V in CH_2Cl_2.

Table 10. Comparison of Half-Wave Potentials for Reduction of OEPFeX and TPPFeX in three solvents containing 0.1 M TBAP[a]

Solvent	Counterion	OEPFe$^+$	TPPFe$^+$	$\Delta E_{1/2}$[b]
CH_2Cl_2	ClO_4^-	0.10	0.22	0.12
	Br^-	−0.34	−0.21	0.13
	Cl^-	−0.42	−0.30	0.12
	N_3^-	−0.52	−0.42	0.10
	F^-	−0.63	−0.50	0.13
DMF	ClO_4^-	−0.18	−0.05	0.13
	Br^-	−0.22	−0.05	0.17
	Cl^-	−0.34	−0.18	0.16
	N_3^-	−0.45	−0.25	0.20
	F^-	−0.64	−0.40	0.24
Py	ClO_4^-	−0.04	0.17	0.21
	Br^-	−0.03	0.17	0.20
	Cl^-	−0.03	0.16(−0.27)[c]	0.19
	N_3^-	−0.03(−0.52)[c]	0.18(−0.33)[c]	0.21
	F^-	−0.04(−0.57)[c]	0.16(−0.54)[c]	0.20

[a] Data taken from K. M. Kadish, L. A. Bottomley, S. Kelly, D. Schaeper, and L. R. Shiue, Reference [169].
[b] $\Delta E_{1/2} = E_{1/2}[TPPFe^+] - E_{1/2}[OEPFe^+]$.
[c] $E_{p,c}$ of second cathodic process.

On changing from the nonbonding solvent CH_2Cl_2 to the weakly coordinating solvent DMF, the cathodic shift in potentials due to X^- decreases and the potential difference $\Delta E_{1/2}$ between the TPPFeX and OEPFeX increases to 0.13–0.24 V, depending on the counterion. It is not clear if the change in $\Delta E_{1/2}$ between CH_2Cl_2 and DMF is due to changes in axial ligation, changes in metal spin state, or changes in iron–porphyrin-plane distance. All three of these are possible. In pyridine, potentials for the reduction of $TPPFe(py)_2^+$ are identical for all counterions and are approximately 0.20 V positive of the $OEPFe(py)_2^+$ potentials. Thus, it appears that when the axial ligand is nonlabile there is a constant different between the OEPFeX and TPPFeX or $OEPFe(L)_2^+$ and $TPPFe(L)_2^+$ potentials, but these values of $\Delta E_{1/2}$ are not identical for all solvents. It is interesting to note that different π-ring systems of iron porphyrins with a given halide may show little difference in $E_{1/2}$, because of the compensating effect of ring-basicity changes and the strength of the bound halide counterion. This appears to be the case of Fe(III) $\rightleftarrows$ Fe(II), where only a 120-mV difference is observed between $(CN)_4$TPPFeCl and TPPFeCl [110]. In contrast, the difference between TPPFeCl reduction and $[(N-CH_3)TPPFeCl]^+$ reduction is 780 mV, with the latter compound reduced at $+0.49$ V [111]. [The actual reaction is that of an Fe(II) oxidation from $(N-CH_3)$TPPFeCl, but because the electrode reaction is reversible, it may be described as occurring in either direction.]

i). Substituent Effects on Half-Wave Potentials

One of the best means of quantifying the effect of electron-donating or electron-withdrawing substituents on the reactions and physical properties of metalloporphyrins is by use of the Hammett linear free-energy relationship [112]

$$\Delta \log K = \log \frac{K^X}{K^H} = \sigma\rho. \tag{22}$$

In this equation σ, a substituent constant, measures the electron-donating or -withdrawing characteristics of X, while the reaction constant ρ measures the sensitivity of the reaction to the electron-donating or -withdrawing characteristics of the substituent. The values of ρ are defined on the basis of the ionization constants of benzoic acid, for which $\rho = 1.0$ is defined for Equation (22).

During the last five years, Equation (22) has been used to calculate substituent effects on shifts of electronic absorption spectra [113], relative spectral intensities [114], and equilibrium constants for addition of axial ligands to the central metal of a number of metalloporphyrins [63, 115–117]. This equation is valid for both equilibrium and rate constants, and—suitably modified—can be applied to measure changes in electrochemical half-wave potentials or heterogeneous electron-transfer rate constants. For the case of half-wave potentials the relevant equation is [118]

$$\Delta E_{1/2} = E_{1/2}^X - E_{1/2}^H = \sigma \rho_{EMF}. \quad (23)$$

In the case of reversible electrochemical reactions, $E_{1/2}$ is related to log K through the Nernst equation, so that Equations (22) and (23) differ by a factor of $2.303RT/nF$ and $\rho_{EMF} = 0.059\rho$ at 25°C for a one-electron transfer.

Investigations of standard tables of σ compiled by several groups [118–121] show that a number of different values of σ exist, depending upon on the polar, resonance, or steric interactions of the substituents. When, in a reaction series, the steric and resonance effects are constant or negligible, shifts of $E_{1/2}$ may be influenced by polar effects only. This has been the case for most correlations with metalloporphyrin reactivity, where σσ has been used almost exclusively.

It should be remembered, when considering Equation (23), that the constant ρ is dependent on the nature of the electroactive group, as well as on the solvent, the supporting electrolyte, and the temperature. Thus, in a given solvent system and at a given temperature, each reactive site (metal or ring) will have a characteristic value of ρ. This has been used, in the case of (p-X)TPPNi [122] and other porphyrins [123] to determine the actual site of the electron transfer. An example of one such plot of $E_{1/2}$ versus σ is given in Figure 12, where the numbers 1 to 4 correspond to the reactions shown in Figure 2. Reaction 5 is that of the anion radical formation and is not shown in Figure 2 (page 170).

The validity of Equation (23) has been demonstrated for thousands of organic and inorganic compounds during the last 25 years [118]. In addition, although the equation is given for reversible reactions, both reversible and irreversible reactions have been studied [118] and have shown good correlations between $E_{1/2}$ and σ. Despite widespread use of linear free-energy relationships, no investigations involving metalloporphyrin redox potentials were published until the mid 1970s. Since then numerous results have appeared in the literature.

Substituent effects on metalloporphyrin redox potentials have been measured principally for three types of porphyrins: the natural porphyrins, substituted tetraphenylporphyrins, and ring-substituted tetraphenylporphyrins. In early measurements of natural-porphyrin redox properties, half-wave potentials were measured by potentiometric methods in the presence of various complexing ligands such as pyridine, picoline, or cyanide [1, 2]. As would be expected, for a given complex, electron-donating substituents on the ring produced cathodic shifts of potential for the reaction Fe(III) ⇌ Fe(II), and electron-withdrawing substituents produced anodic shifts of potential for the same reaction. However, quantitative shifts of the potential using a linear free-energy relationship were not possible, due to an inexact knowledge of the appropriate type of σ to be selected. It was not until the synthesis of

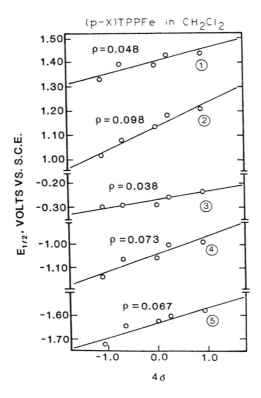

Figure 12 Plot of $E_{1/2}$ versus σ for five of the electrode reactions of (p-X)TPPFeCl in CH_2Cl_2, 0.1 M TBAP. The numbers 1 to 5 correspond to the peaks shown in Figure 2.

symmetrically substituted porphyrins such as (p-X)TPPH$_2$ that quantitative correlations could be obtained.

The first systematic studies of substituent effects on metalloporphyrin reactivity began to appear in 1975. Working independently, the groups of Kadish and Davis in the U.S.A. and Gross in France investigated half-wave potentials and electron-transfer rate constants for a series of different metalloporphyrins. Essentially identical conclusions were obtained by each laboratory, although the magnitude of ρ was different, due to the type of porphyrin selected. Studies by Kadish et al. [50, 59, 122–125] included mainly electrode reactions of phenyl-substituted tetraphenylporphyrins (p-X)TPPM, while those of Gross et al. [110, 126–128] concentrated on electrode reactions of tetraphenylporphyrins where the substituent was placed directly on the porphyrin ring. Synthesis of these pyrrole-substituted porphyrins has been described in detail by Callot [110, 129, 130].

As of this writing only a few studies of iron-porphyrin substituent effects have been published. The results obtained are generally as expected. However, in order to put the data in perspective it is necessary to first consider the results which have been published for substituent effects on a number of other metalloporphyrin complexes.

The initial study of (p-X)TPPH$_2$ redox reactions [124] was carried out in a variety of nonaqueous solvents. Four electrode reactions were observed, of which all gave linear plots of $E_{1/2}$ versus σ. Values of ρ were found to be a function of solvent, as is predicted. For example, ρ for reduction of (p-X)TPPH$_2$ to yield the π-anion radical [(p-X)TPPH$_2$]$^-$ decreased from 0.073 V in CH$_2$Cl$_2$ to 0.053 V in DMSO.

Later studies of substituent effects involving (p-X)TPPM in CH$_2$Cl$_2$ included complexes where the central metal was Co [59], Mn [122], Ni [122], Fe [50], and Ru [131] as well as VO [123], Cu [123] and Zn [123]. A summary of substituent effects for the ring-centered reactions of several different metalloporphyrin complexes indicated that the average ρ for ring oxidation (to yield either cation radicals or dications) was 0.07 ± 0.01 V. A similar value of 0.06 ± 0.01 V was reported for the first ring reduction to yield π-anion radicals, and 0.07 ± 0.01 V for the second ring reduction to yield dianions [123]. Earlier, a value of 0.08 V was reported by Wolberg [47] for several complexes.

Table 11 summarizes values of ρ obtained in CH$_2$Cl$_2$ for Fe(III)/Fe(II) as well as other complexes of (p-X)TPPM. From the data it appears that, for these specific metalloporphyrins, the value of ρ for ring-centered reactions is fairly independent of the charge on the reacting species or the nature of the central metal. This was true for central metals of oxidation state +1, +2, +3, or +4.

The constancy of ρ over a wide range of central metal ions is a good indication that electron transfer involves the porphyrin ring and not the metal ion. Conversely, deviations of ρ from that which is expected might be a good indication that the electrode reaction involves an orbital of the central metal and not the ring. In these cases ESR and optical methods can be used to further investigate the electrode reaction.

For a series of transition metals complexes of (p-X)TPP^{-2} it was found that ρ for the electrode reaction M(III) ⇌ (M(II) apparently depended on the number of d-electrons of the metal [122]. As the number of electrons increased from Mn(II) to Fe(II) to Ni(II), ρ decreased from 0.054 to 0.038 to 0.018 V in CH$_2$Cl$_2$. The last, low value for Ni(II) ⇌ Ni(III) was used [122] to differentiate the overlapping metal-centered reaction of TPPNi [132] from that of the π-cation radical (for which ρ = 0.089V).

A number of results similar to those for complexes of (p-X)TPP^{-2} have been observed for reduction or oxidation of ring-substituted tetraphenylporphyrins [126–128]. However, for these complexes, the placement of the substituent directly on the porphyrin ring rather than on the phenyl group

Table 11. Values of ρ Calculated According to Equation (23) for Several Complexes of $(p\text{-}X)$TPPM and $(X)_n$TPPM

	$(p\text{-}X)$TPPM[a]			$(X)_n$TPPM[b]	
Central metal	Cation radical	M(III) ⇌ M(II)	Anion radical	Cation radical[c]	M(III) ⇌ M(II)[c]	Anion radical
Mn	0.07	0.05	—	—	0.02	0.27[d]
Fe	0.10	0.04	0.07	0.15	0.01	0.29[d]
Co	0.09	0.03	—	0.08	—	0.27[d]
Ni	0.09	0.02	0.05	—	—	0.27[d]
Cu	0.08	NR	0.06	0.14	NR	0.29
Zn	0.06	NR	0.05	0.12	NR	0.28[d]
2H	0.06	NR	0.07	0.15	NR	0.29

[a] Data taken from K. M. Kadish and M. M. Morrison, References [122] and [123].
[b] Data calculated from A. Giraudeau, H. J. Callot, J. Jordon, I. Ezahr, and M. Gross [110], and A. Giraudeau, H. J. Callot, and M. Gross [128].
[c] Calculated from potentials of TPPM and $(CN)_4$TPPM using σ.
[d] Calculated from potentials of TPPM and $(CN)_x$TPPM using σ^-.

results in much larger shifts of potential, for which more exact correlations between $E_{1/2}$ and σ can be investigated.

In all solvents, reduction of $(CN)_X$TPPM or $(Br)_X$TPPM where $X = 1, 2, 3$, or 4 results in an anodic shift of potential with an increase in the electron-withdrawing nature of the substituent. Analysis of the potential shifts by Equation (23) leads to $\rho = 0.28$ V [127, 128]. This value of ρ is approximately four times that observed for the para-substituted tetraphenylporphyrin complexes (p-X)TPPH$_2$, and was identical in CH$_2$Cl$_2$ [128] and in DMF [127].

Similar cathodic shifts were observed for the ring reductions of complexes of $(CN)_X$TPPM, where M was Mn(III) [110], Fe(III) [110], Co(III) [110], Cu(II) [110, 126–128], or Ag(II) [127]. For each complex the shift in porphyrin-ring potential was approximately 0.24 to 0.27 V per cyano group. Thus, for the case of Fe(I) reduction to yield the anion radical, a potential of -1.700 V is reported starting with TPPFeCl, and -0.730 V starting with $(CN)_4$TPPFeCl. Based on these shifts of potential, approximate values of ρ were calculated using Equation (23) and are listed in Table 11 for comparison with the complexes of (p-X)TPPM.

As already mentioned, placement of substituents directly on the porphyrin ring produces much larger shifts of potential than tetraphenyl substitution. Because the precision of a given measurement is 5 mV, and shifts of up to 1000 mV are observed, analysis of linear free-energy relationships may be examined in more detail for ring-substituted complexes than for the complexes of (p-X)TPPM.

In the initial studies by Gross et al. [126, 127], σ was selected for use in Equation (23). However, in a latter study [128] the authors realized that the earlier selection of σ was in error, and elected to use σ$^-$ for reductions. Use of this substituent constant resulted in a much better fit of the plot to the data. It is interesting to note that identical values of $\rho = 0.28$ V are obtained for reduction of (X)TPPH$_2$ and (X)TPPCu, independent of whether the substituent constant used was σ or σ$^-$. However, this may be fortuitous. The first study was performed in DMF and the second in CH$_2$Cl$_2$, and because of solvent effects on ρ, direct comparisons of the magnitude of ρ should not be made.

An interesting result arising from the electrochemical studies of polysubstituted porphyrins is that there appears to be a clear-cut difference between the oxidation and reduction behavior of a series of ring-substituted porphyrins and that the site of electron transfer (either addition or abstraction of one electron) may not be the same in both reactions. Based on fits of half-wave potential to linear free-energy plots, it was noted that the reduction potentials correlated with σ$^-$ while the oxidation potentials correlated with σ. Based on this difference (as well as the much smaller shift of $E_{1/2}$ with the number of cyano groups for oxidation), it was concluded that reductive electron transfer was to the conjugated π-system in which the β-substituents were in direct resonant interaction. In contrast, it was concluded that abstraction of an electron from the porphyrin was not from the conjugated π-system but rather from the lone electron pairs found on the pyrrolic nitrogens. The smaller shift of $E_{1/2}$ for oxidation is thus explained on the basis that interactions of the four nitrogens with ring substituents is mostly inductive and inductive interactions are weaker than conjugative (resonant) ones [133]. This interpretation is in agreement with the conclusions of Stanienda and Biebl [4, 134] on the electrochemical reactivity of metalloporphyrins. However, these results do not contradict the results obtained for para-substituted tetraphenylporphyrins. In these latter complexes of (p-X)TPPM, the substituents are substituted on the four phenyl groups which are not coplanar with the phenyl ring [135, 136]. In these cases, linear correlations are obtained only if the nonresonant parameter σ is used for both oxidations and reductions [50, 59, 122–125, 137].

For all identified ring-centered reactions the shift of $E_{1/2}$ was 0.24 to 0.27 V per cyano group. For the metal-centered reaction Mn(III) $\rightleftarrows$ Mn(II) this difference was 0.052 V per cyano group, while for the metal-centered reaction Fe(III) $\rightleftarrows$ Fe(II) it was only 0.01. Clearly, the decrease in substituent effects for metal-centered reactions is far less than those for the ring centered reactions, and this information can be used, similarly to the case of (p-X)TPP^{-2} complexes [122], to differentiate metal-centered from ring-centered reactions. It is unfortunate that the electrode reaction Fe(III) $\rightleftarrows$ Fe(II) could not be shifted to extremely positive potentials by the use of ring substituents alone. From previous studies, a potential of 0.3 to 0.5 V is predicted, but this could not be obtained.

As a final point, it is interesting to note that the reactions Fe(II) $\rightleftarrows$ Fe(I) give $E_{1/2}$ values which shift for each cyano group by 0.20 V. This is in line with what would be expected from the data on the (p-X)TPPM(III) complexes and suggests there is another effect operating on shifting the electrode reactions of Fe(III) and Fe(II). This most likely is the increased strength in the Fe(III)-chloride bond, which does not exist for Fe(II). On the other hand, if one can make the statement that metal-centered reactions should exhibit smaller substituent effects than those of the ring, this might imply, for the reactions Fe(II)/Fe(I), that there exists some radical character to the reduced species.

Prior to the work of Kadish and of Gross most studies of substituent effects on redox processes involved comparing differences in potential between the reactions Fe(III)/Fe(II) for iron complexes of natural porphyrins, or differences between the metal-centered reactions of the synthetic porphyrins OEPFeCl and TPPFeCl. For a series of natural porphyrins in aqueous media containing pyridine the most positive reduction potential was usually found in iron protoporphyrin IX, while the most negative was usually in iron mesoporphyrin or iron coproporphyrin [1]. However, the difference in E^0 was usually less than 100 mV. For example, with pyridine as an axial ligand at pH 9.6 the Fe(III)/Fe(II) potential was +0.015 and −0.063 V for protoporphyrin

IX and mesoporphyrin, respectively. Likewise with cyanide as axial ligand E^0 for reduction of Fe(III) differed by only 0.064 V between iron complexes of protoporphyrin and coproporphyrin. In this case the potentials were -0.183 V and -0.247 V, respectively [1]. Similar small differences were found by Kadish and Davis for several OH^- complexes of natural porphyrins in DMSO [48].

Larger differences are found between the potentials of metal complexes containing OEP^{-2} and TPP^{-2}, although (as already seen in Table 10) this depends on the nature of the counterion, as well as the metal. In DMSO potentials of $OEPH_2$ are shifted cathodically from $TPPH_2$ by $\simeq 0.40$ V for the formation of π-anion radical and by 0.15 V for the formation of the π-cation radical in CH_2Cl_2. Likewise, for a series of similar complexes containing central metals, the difference in potential for ring reduction between the more basic OEPM and the TPPM complex is approximately 0.27 V, while shifts in potential for ring oxidation are substantially less and average about 0.09 V. These numbers must not be taken as absolute, since they are taken from data on approximately 20 different complexes obtained under slightly different experimental conditions [138, 139]. However, despite the fact that the data were taken in different laboratories, clear-cut differences are evident between the "substituent effects" for oxidation and those for reduction.

In order to obtain comparisons between naturally occurring and synthetic iron complexes, Kadish and Larson [140] measured half-wave potentials for a series of complexes in DMF under the same experimental conditions. The results of this study are summarized in Table 12 along with similar data for $(CN)_4$TPPFeCl. Neglecting this latter complex, the most positive potential is always that for the reactions of TPPFeCl and the most negative is for reduction of OEPFeCl or EtioFeCl. This was true for all three electroreductions of each complex. The potential difference between OEPFeCl and TPPFeCl was 160 mV for the reaction Fe(III)/Fe(II), 210 mV for the reaction Fe(II)/Fe(I), and 250 mV for the reaction Fe(I)/Fe(I) radical. Thus, as we would predict, the largest effect is observed for reactions involving the π-ring system, and the smallest for those involving the central metal.

Since absolute porphyrin-ring basicities are not known for each of the complexes investigated, an arbitrary scale of ring basicity was constructed and

Table 12. Half-Wave Potentials for Reduction of Ferric Porphyrins in DMF[a]

Complex	Fe(III)/Fe(II)	Fe(II)/Fe(I)	Fe(I)/Fe(I) Radical	Reference
$(CN)_4$TPPFeCl	-0.12	-0.26	-0.73	[110]
TPPFeCl	-0.18	-1.03	-1.65	[9, 20, 50, 110]
ProtoFeCl	-0.27	-1.19	-1.68	[48, 49, 140]
DeuteroFeCl	-0.30	-1.20	-1.79	[10, 140]
EtioFeCl	-0.34	-1.25	-1.91	[140]
OEPFeCl	-0.34	-1.24	-1.90	[140]

[a] Potentials are reported as $E_{1/2}$ (V) versus SCE.

$E_{1/2}$ plotted for each compound versus this basicity. An arbitrary slope of 1.0 was assigned to the electrode reaction involving formation of the π-anion radical from Fe(I). Using this slope of 1.0, a slope of 0.88 was obtained for the reaction Fe(II)/Fe(I), and 0.53 for the reaction Fe(III)/Fe(II). Thus, from these data it is seen that the effects of ring substituents in DMF are about 160 to 180% greater for reactions involving the π-system then for reactions involving Fe(III) $\rightleftarrows$ Fe(II). This increase agrees with a similar increase measured in ρ for the same reaction of TPPFeCl in CH_2Cl_2, DMF, and DMSO [50].

ii). Measurements of Electron-Transfer Rates

Electron-transfer rate constants of metalloporphyrins have been measured for Fe(III)/Fe(II), Fe(II)/Fe(I), and oxidation of Fe(III). It has been postulated by Kadish and Davis [48] and by Constant and Davis [93] that slow electron-transfer rates will be obtained for the Fe(III) $\rightleftarrows$ Fe(II) reaction whenever a change of spin state occurs.

When considering all possible spin states of each reactant and product (see Figures 8 and 9), nine types of electron transfer are possible. There are high-spin Fe(III) reduced to high-, intermediate-, or low-spin Fe(II); intermediate-spin Fe(III) reduced to high-, intermediate-, or low-spin Fe(II); and low-spin Fe(III) reduced to high-, intermediate-, or low-spin Fe(II). Five of these nine possibilities may be eliminated on the basis of the known chemistry of iron-porphyrin systems. If we assume, for the moment, that electron transfer between redox species of different spin multiplicities is spin-forbidden [141], or at least spin-restricted, we can exclude one of the remaining four, which is high-spin Fe(III) reduced to intermediate-spin Fe(II). This leaves the following three electrode reactions which may exist in the iron-porphyrin system:

$$Fe(III)_{HS} \rightleftarrows Fe(II)_{HS}, \qquad (24)$$

$$Fe(III)_{IS} \rightleftarrows Fe(II)_{IS}, \qquad (25)$$

$$Fe(III)_{LS} \rightleftarrows Fe(II)_{LS}, \qquad (26)$$

where HS, IS, and LS represent the high-, intermediate-, and low-spin forms of the iron respectively. One question that has been considered in the literature but perhaps not yet answered is how the spin states of iron(III) and/or iron(II) affect the heterogeneous electron-transfer rate constant, and whether the effect of spin state can be separated from that of axial ligation and changes in metal–porphyrin-plane distance. The difficulty in answering these questions is that the data on spin states which are available in the literature were not measured under the same experimental conditions as those in which the electron-transfer rate constants were calculated. In addition, there is generally a change of axial ligation associated with the change of spin state, and this may be the determining factor in whether k^0 is rapid or slow. However, despite these obvious limitations, several conclusions have been drawn regarding spin state and k^0.

In the first study of Fe(III)/Fe(II) porphyrin rate constants it was reported that the reduction of low-spin Fe(III) dicyanoprotoporphyrin IX to low-spin Fe(II) dicyanoprotoporphyrin (IX) had a rate constant of 4.0 cm/sec [142]. In contrast, reduction of ferriheme monomer in 50% ethanol-water mixtures has a rate constant of 0.2 cm/sec [143]. In this latter complex the iron is most likely high-spin in both the oxidized and the reduced form of the complex. However, it should be noted that x-ray data [144] show changes in the iron–porphyrin-plane distances upon reduction, and this may also be responsible for the slowing of the rates.

Rate-constant measurements as a function of complexed axial ligand have also been undertaken. In the first published study of porphyrin Fe(III)/Fe(II) rates by cyclic voltammetry, Kadish and Davis found that TPPFeCl in DMSO had a rate constant of 2×10^{-3} cm/sec and that reduction of the same complex in DMF yielded a rate constant of 8×10^{-4} cm/sec [48]. This decrease in rate was attributed to the fact that in DMF, Cl^- was associated and the reactant was TPPFeCl [or TPPFeCl(DMF)], while in DMSO, Cl^- was dissociated and the reactant was TPPFe(DMSO)$_2^+$. This would imply that loss of a ligand upon reduction slows the rate of electron transfer. A similar trend has been reported for reductions of Co(III) ⇌ Co(II) [145].

It has been shown that electrochemical rate constants for the reaction Fe(III)/Fe(II) are proportional to the base strength of the axially complexed ligand [93] provided that both the oxidized and reduced forms of the species are bis coordinated. In contrast, however, iron porphyrins complexed by poly-L-histidine chains in aqueous solutions have been shown to be completely nonelectroactive [146].

Studies of rates in DMF-water mixtures have shown that changing from DMF to water as a pure solvent increases the rate constant by a factor of 100 [147]. In this case the authors propose specific interactions between the solvent and the periphery of the porphyrin-ring system, and that the actual path of electron transfer is via the conjugated porphyrin ring. This postulate has been made by others [148] and is the conclusion of several independent studies.

In order to observe how k^0 changes with changes in the base strength of the complexed axial ligand, rate constants were measured for the reduction of a series of (p-X)TPPFeCl complexes in butyronitrile [50]. In this study, complexes containing electron-donating substituents (which produced a higher electron density on the ring) possessed substantially increased k^0's for the metal-centered reactions. This agrees, at least qualitatively, with data obtained for the reduction of hemin in DMA [93]. In this latter study an increase in k^0 for the Fe(III) ⇌ Fe(II) reaction was observed with an increase in the pK_a of the bound ligand. In contrast, however, rate studies of (p-X)TPPFeCl* as

---

* Bottomley, L. A., and Kadish, K. M., unpublished results.

well as other porphyrins [140] in DMF have shown the opposite trend, that is, a decrease in k^0 with increasing negative charge on the porphyrin ring. In the latter study the porphyrins utilized were TPPFeCl, iron protoporphyrin IX, iron deuteroporphyrin, OEPFeCl, and EtioFeCl.

Despite the apparent conflict of data, it is hoped that studies of heterogeneous electron-transfer rate constants will shed light on the actual path of electron transfer to and from the central iron atom. There have been a number of hypothesis concerning the manner in which electrons are passed to and from the iron in cytochrome c [149], hemoglobin [89], and synthetic porphyrins [93, 105, 143, 146, 148, 150, 151]. Claims have been made [148] that *homogeneous* electron-transfer paths to and from the central metal are via the conjugated π-system of the porphyrin plane or by the groups occupying the fifth and sixth coordination positions of the metal. Separate binding sites on cytochrome c have been taken as evidence that two distinct pathways do exist in the molecule [152, 153]. This has not been observed directly by electrochemical methods, but might be implied from the two possible electron transfer mechanisms for reduction and oxidation of Fe(III) complexes (see following sections).

G. OXIDATION-REDUCTION MECHANISMS

During the last five years a number of electrochemical studies have been undertaken in which the objective has been to determine the exact mechanism of electron transfer. Based on these studies, a clear picture is now beginning to emerge as to how the solvent may direct the path of electron transfer from Fe(III) to Fe(II) and back again to Fe(III) for the case of "reversible" electron transfers. Electron transfer of Fe(III)/Fe(II) has now been investigated in over fifteen different nonaqueous solvents. In this review, mechanisms in three commonly used solvents will be discussed.

i). Reduction Mechanism in DMF

A number of electrochemical studies of iron porphyrins have been undertaken in DMF [9, 20, 43, 63, 117, 140, 147]. As already mentioned, this solvent was usually chosen because of its good solvating properties for porphyrins and because of its wide potential window, especially for reductions. However, despite reversible waves in many cases, the exact nature of the DMF-porphyrin interaction was not well understood. Kadish and Larson [140] had proposed the existence of a transient $[PFe(III)Cl]^-$ upon reduction of Fe(III) to Fe(II):

$$PFe(III)Cl \underset{\text{rate-controlling}}{\overset{e}{\rightleftarrows}} [PFe(III)Cl]^- \rightarrow PFe(II) + Cl^-. \qquad (27)$$

This mechanism was based on changes in heterogeneous rates of electron transfer and was not substantiated.

A comparison of the Fe(III) reduction potentials measured in DMF (Tables 5 and 10) indicates that $E_{1/2}$ is influenced by the nature of X, although to a lesser extent than in CH_2Cl_2 or $EtCl_2$. The difference in potential between reduction of TPPFeX when $X = ClO_4^-$ and $X = F^-$ is substantially diminished from the difference observed in nonbonding media (350 mV in DMF compared to 710 mV in $EtCl_2$). This decrease could be attributed to an increase in the relative stabilization of Fe(II) versus Fe(III) from one solvent to another or by changes in the stoichiometry or spin state of the Fe(III) reactants. Brault and Rougee [62] have reported that TPPFe(II) coordinates only one DMF molecule axially and does so very weakly (formation constant 6.2 ± 0.3 when measured in benzene). Because this axial coordination of Fe(II) is so weak, the relative stabilization of Fe(II) over Fe(III) is not a plausible explanation for the observed difference in reduction potentials between measurements made with DMF as solvent and those made in $EtCl_2$.

When $X = ClO_4^-$ and Br^-, an identical $E_{1/2}$ of -0.05 V is observed. This implies that the counterion has been displaced by solvent molecules as follows for $X = ClO_4^-$:

$$TPPFeClO_4 + 2DMF \rightleftharpoons TPPFe(DMF)_2^+ + ClO_4^-. \qquad (28)$$

Verification that DMF displaces ClO_4^- and Br^- was obtained by an electrochemical titration of $TPPFeClO_4$ in CH_2Cl_2. Nerstian behavior was observed for the Fe(III)/Fe(II) and Fe(II)/Fe(I) redox couples between 1.00 mM and 1.00 M DMF. Analysis of the peak shifts as a function of DMF concentration* indicated that Fe(III) was complexed by two molecules of DMF at high DMF concentrations and no DMF molecules at low concentration. At the same time TPPFe(II) was axially complexed by only a single DMF molecule at high concentration. The formation constants for addition of two DMF molecules to Fe(III) according to the reaction (28) and one DMF molecule to Fe(II) were calculated as 9×10^4 and 2.0, respectively.*

The observed potential for the Fe(III)/Fe(II) couple when Cl^- is the $TPPFe^+$ counterion is midway between $X = ClO_4^-$ and F^- (see Table 5). Association of Cl^- to $TPPFe^+$ under the experimental conditions of the electrochemical measurements was verified spectrophotometrically, while dissociation of the Cl^- at lower (TPPFeCl) was confirmed from conductance studies [46].

Figure 13 shows a cyclic voltammogram obtained for the Fe(III)/Fe(II) couple of TPPFeCl in DMF. On the negative potential sweep, only one cathodic process is observed, while on the reverse sweep, one or two peaks are observed, depending upon the rate of potential scan. At slow scans, one broadened anodic process is observed. Addition of Cl^- in the form of TBACl causes a decrease in peak width. On the other hand, in the absence of excess

---

* Kadish, K. M., and Bottomley, L. A., unpublished results.

Oxidation-Reduction

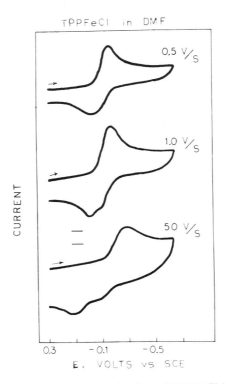

Figure 13 Cyclic voltammogram for reduction of TPPFeCl in DMF, 0.1 M TBAP. (Reprinted with permission of the American Chemical Society [46].)

Cl^-, as the scan rate is increased the one broad anodic process splits into two processes. The oxidation peak at more positive potentials is essentially identical to that obtained when the starting material is $TPPFeClO_4$. Current-voltage curves similar to those depicted for Cl^- were also obtained when $X = N_3^-$ and F^-. Several such curves for $TPPFeN_3$ are shown in Figure 14.

Nicholson-Shain [15] strategies and variable-temperature electrochemistry may be employed to evaluate any equilibrium in solution and to detect possible reactions coupled to the electron transfer. This strategy includes variation in potential scan rates and measurements of both peak potentials and peak currents as a function of scan rate. Such a plot of cathodic peak potential dependence on scan rate is depicted in Figure 15. Experimentally obtained slopes of -30 mV cathodic shift with tenfold increase in scan rate for TPPFeX where $X = Cl^-$, N_3^-, or F^- agree with the predicted values [15] for the case of a rapid chemical reaction following the electron transfer. When $X = ClO_4^-$, or Br^-, the reactant is $TPPFe(DMF)_2^+$ and reversible electron transfers without coupled chemical reactions are obtained. When $X = Cl^-$, N_3^-, or F^-, the reactant is either TPPFeX or TPPFeX(DMF), but the exact stoichiometry cannot be unambiguously assigned from the available spectrophotometric

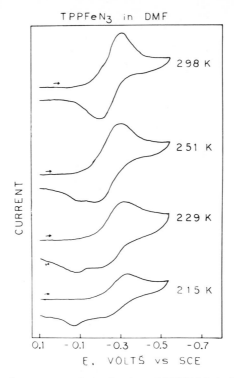

Figure 14 Cyclic voltammogram for reduction of TPPFeN$_3$ in DMF, 0.1 M TBAP.

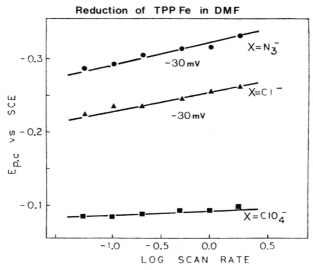

Figure 15 Plots of $E_{1/2}$ versus scan rate showing characteristic shift of $-(30$ mV$)/\log v$ for an EC mechanism.

Oxidation-Reduction

and electrochemical data. In either case, the reduction step generates [TPPFeX]⁻, which readily dissociates to form the ultimate product, TPPFe(DMF).

The fact that TPPFe(DMF) is the ultimate Fe(II) product in the reduction is also shown from the variable-temperature electrochemistry of TPPFeX complexes in DMF. An example is given in Figure 14 for the reduction and reoxidation of TPPFeN₃ in DMF. The overall reduction process is TPPFeN₃ ⇌ TPPFe(DMF). For the reoxidation at 298 K, TPPFe(DMF) converts to [TPPFeN₃]⁻ before electron transfer and a reversible C-V curve is obtained. Lowering of the temperature slows reassociation of the anion, and direct oxidation of TPPFe(DMF) proceeds at $E_{p,a} = -0.08$ V, a potential almost identical to that observed for TPPFeClO₄ in the same solvent. At 215 K reoxidation via [TPPFeN₃]⁻ has decreased to a negligible amount and almost all oxidation is from TPPFe(DMF) directly to [TPPFe(DMF)₂]⁺.

All of the complexes of TPPFeX which have been investigated to date have been shown to follow the same mechanism, which is illustrated in scheme II below [46]:

Scheme II:
$$\begin{array}{ccc} \text{TPPFeX} & \xrightarrow{e^-} & [\text{TPPFeX}]^- \\ X \updownarrow 2\,\text{DMF} & & X \updownarrow \text{DMF} \\ [\text{TPPFe(DMF)}_2]^+ & \xrightarrow[\text{DMF}]{e^-} & \text{TPPFe(DMF)} \end{array}$$

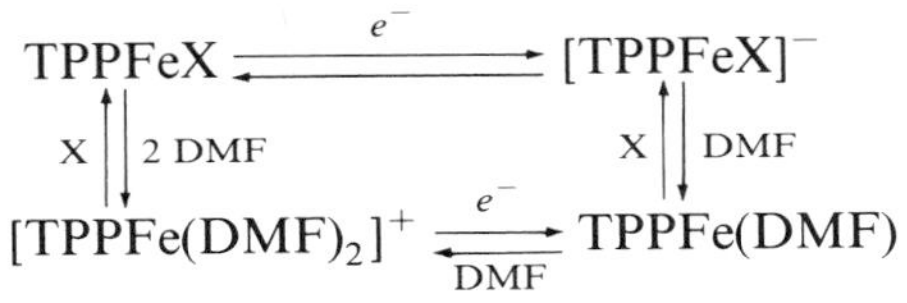

At rapid scan rates or low temperature, oxidation of TPPFe(DMF) can proceed via two different pathways, i.e., directly to [TPPFe(DMF)₂]⁺ or via [TPPFeX]⁻ to TPPFeX. Each of these processes has a characteristic half-wave potential and accounts for the two anodic peaks observed at high scan rate and low temperature. At low scan rate or high temperature, however, TPPFe(DMF) has time to interconvert to [TPPFeX]⁻ prior to the electron transfer and is oxidized by the thermodynamically easier path (more negative potential). Likewise, addition of free X shifts the equilibrium toward [TPPFeX]⁻. This chemical reaction causes the coalescence of the two anodic processes observed at high scan rates.

ii). Reduction Mechanism in Me₂SO

Typical cyclic voltammograms of TPPFeX complexes dissolved in Me₂SO are depicted in Figure 16. For X = ClO₄⁻, Br⁻, or Cl⁻, superimposable cyclic voltammograms were obtained for each of the three reduction steps. The electrode reaction occurring at -0.09 V corresponded to the Fe(III)/Fe(II) redox couple, at -1.14 V to the Fe(II)/Fe(I) redox couple, and at -1.67 V to the formation of the anion radical [117]. The superimposability of these voltammograms implies identical reactants and products at the electrode surface, which are TPPFe(Me₂SO)₂⁺ and TPPFe(Me₂SO)₂. This was observed

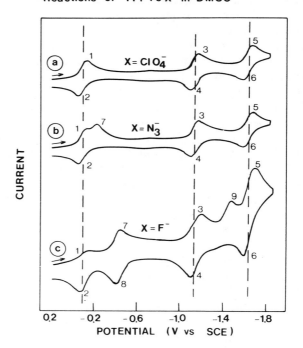

Figure 16 Cyclic voltammogram of several TPPFeX complexes in DMSO, 0.1 M TBAP: (a) TPPFeClO$_4$, (b) TPPFeN$_3$, and (c) TPPFeF.

for similar porphyrins by spectroscopic and conductometric results [154] which indicate dissociation of the counterion in this solvent system.

In contrast to complexes where X = ClO$_4^-$, Br$^-$, or Cl$^-$, the voltammogram of TPPFeN$_3$ dissolved in Me$_2$SO [Figure 16(b)] has two cathodic peaks at -0.13 and -0.25 V, and indicates the presence of more than one Fe(III) form in solution. The first reduction peak is at an identical potential to that of the bis Me$_2$SO adduct, while the second more cathodic peak is at a potential suggesting axial complexation of TPPFe(Me$_2$SO)$_2^+$ with N$_3^-$. All other oxidation and reduction potentials are identical to those for complexes of TPPFeX where X = Cl$^-$, Br$^-$, or ClO$_4^-$ [Figure 16(a)].

Complexes of TPPFeF are also able to exist undissociated in DMSO, as evidenced by both the electrochemical data and the electronic absorption spectra [46]. Comparison of the visible spectrum of TPPFeF in Me$_2$SO with that reported by Cohen [155] for both the α and the β form of TPPFeF dissolved in CH$_2$Cl$_2$ shows striking similarities. Cohen reports absorption maxima at 645, 610, and 550 nm. The spectrum obtained under electrochemical conditions [46] has a maxima at 649 and 610 and a shoulder at 544 nm.

Oxidation-Reduction

Using the 610-nm peaks as the reference, Cohen reports the relative peak absorptivities of 0.87 and 0.92 for the 645- and 550-nm peaks, respectively. The relative peak absorptivities observed in Me_2SO are 1.28 and 0.86 for the 649- and 544-nm peaks, respectively. This implies a substantial interaction by the solvent, which can be explained by direct axial coordination of the Fe center by a Me_2SO molecule.

The most convincing evidence that F^- is associated in DMSO comes from the cyclic voltammogram [shown in Figure 16(c)]. Five separate electrode reactions were observed at $E_{1/2} = -0.09, -0.40, -1.15, -1.39$, and -1.63 V, respectively. The redox couples at $-0.09, -1.15$, and -1.63 V (labeled peaks 1 to 6) are identical to potentials observed for reduction of TPPFeCl, TPPFeBr, TPPFeClO$_4$, and TPPFeN$_3$. The reactions at $E_{1/2} = -0.40$ V (peaks 7 and 8) and -1.39 V (peak 9) correspond to reductions of Fe(III) and Fe(II) complexes which are axially complexed by F^- ion. Evidence for a mono fluoride complex of Fe(II) comes both from electrochemical titrations of TPPFe and TPPFe(Me$_2$SO)$_2$ with F^- and from the 240-mV cathodic shift in polarographic half-wave potential for the reaction Fe(II) $\rightleftarrows$ Fe(I) when compared with TPPFe(Me$_2$SO)$_2$ [46].

Finally, evidence has also been obtained for the formation of TPPFeF$_2^-$ adducts in Me$_2$SO (and CH$_2$Cl$_2$), and its electrochemical reactivity recorded. The bis fluoride complex may be obtained from titrations of TPPFeClO$_4$ with F^- [46] or can be synthesized directly.* Reduction of TPPFeF$_2^-$ yields initially the five-coordinated species [TPPFeF]$^-$, whose spectrum is identical to that observed in CH$_2$Cl$_2$.

Based on both electrochemical titrations and spectral characterization, the following mechanism is proposed for reduction of TPPFeX in Me$_2$SO:

Scheme III:

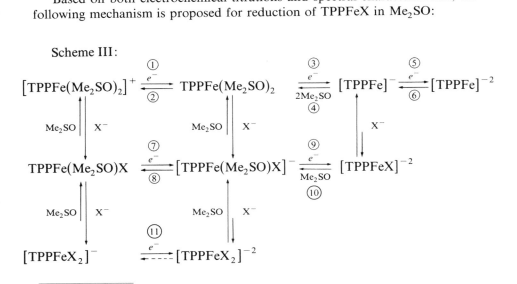

---

* Marchon, J. C., private communication.

For complexes of TPPFeX where X = ClO_4^-, Br^-, or Cl^-, the reduction-oxidation process is via the top sequence, steps 1–6. When X = N_3^- or F^-, however, the primary reactant at the electrode is TPPFe(Me$_2$SO)X, which is in equilibrium with TPPFe(Me$_2$SO)$_2^+$. For the case of X = F^- the former species is reduced to [TPPFe(Me$_2$SO)F]$^-$ (peak 7), which exists in equilibrium with TPPFe(Me$_2$SO)$_2$. Reduction to the Fe(I) species produces the uncomplexed [TPPFe]$^-$, even though a transient [TPPFeX]$^{-2}$ species is indicated for X = F^-. For all complexes, the anion-radical formation (peaks 5 and 6) is unaffected by counterion. Finally, for the single case where X = F^- and at high molar ratios of F^- to porphyrin, the bis F^- adduct of Fe(III), [TPPFeF$_2$]$^-$, is predominant in solution.

The electrochemical reduction of TPPFeX has also been investigated in mixed solvent systems containing Me$_2$SO and pyridine. In addition pyridine titrations of [TPPFe(Me$_2$SO)]$^+$ have been followed electrochemically [117]. Pyridine will not displace a Me$_2$SO molecule complexed to Fe(III) even at concentrations greater than 1.0 M. In contrast, pyridine will replace both solvent molecules on Fe(II) with stepwise formation constants ranging from log K_1 = 1.85 to 2.44 and log K_2 = 0.80 to 1.29. It is interesting to note that $K_2 < K_1$ for this reaction, implying no change of spin state upon changing axial ligands [117].

In the presence of high pyridine concentration, the electrochemical reactions are a combination of those determined for the bis pyridine Fe(II) complex TPPFe(py)$_2$ and the bis Me$_2$SO Fe(III) complex, TPPFe(Me$_2$SO)$_2^+$ [51, 63, 117]. The overall reactant and products have been determined at several different Me$_2$SO pyridine mixtures, and the electrode mechanism is given by scheme IV below:

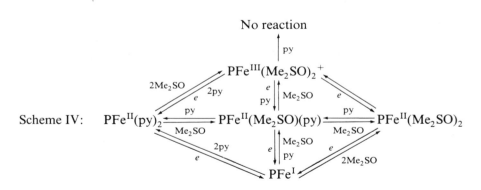

iii. Reduction Mechanism in CH$_2$Cl$_2$ and EtCl$_2$

Three different pathways for electron transfer may be observed in CH$_2$Cl$_2$ (or EtCl$_2$) as a function of the counterion bound to the iron center [46]. When

$X = ClO_4^-$, the reduction is straightforward and yields TPPFe at potentials of +0.22 to +0.24 (see Table 5). When $X = N_3^-$ or F^-, the product is $[TPPFeN_3]^-$ or $[TPPFeF]^-$, which is formed at an $E_{1/2}$ of -0.38 or -0.47 V respectively in $EtCl_2$. The reduction product, $[TPPFeX]^-$, is not a transient intermediate, and discrete five-coordinated Fe(II) complexes may be observed [46, 156]. This has been discussed in detail for the case of TPPFeF. When $X = Cl^-$ or Br^-, coupled chemical reactions are observed and the reduction-reoxidation sequence involves two separate oxidation-reduction pathways [46]. This is shown in scheme V, which summarizes the oxidation-reduction pathway for five different counterions:

Scheme V:

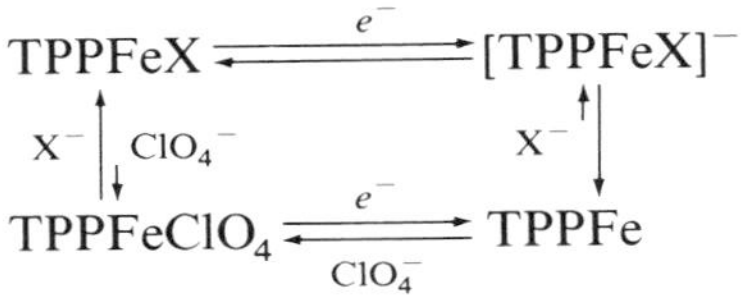

$$\begin{array}{ccc} TPPFeX & \xrightleftharpoons{e^-} & [TPPFeX]^- \\ X^- \Big\downarrow ClO_4^- & & X^- \Big\uparrow \\ TPPFeClO_4 & \xrightleftharpoons[ClO_4^-]{e^-} & TPPFe \end{array}$$

When $X = ClO_4^-$ the bottom pathways is followed. This is the easier reduction in the series and $E_{1/2}$ for Fe(III)/Fe(II) is determined by the stability of the $TPPFe^+ - ClO_4^-$ interaction. When $X = N_3^-$ or F^-, the top electron-transfer path in scheme V is followed. No appreciable dissociation of $[TPPFeX]^-$ occurs, and reversible electron transfers are obtained. The half-wave potentials for these reactions are the most cathodic of the series, and the reductions are the most difficult. It is significant to note that for these complexes, the half-wave potentials are influenced by the relative stabilities of the Fe(III)- *and* Fe(II)-halide interactions. Increases in the Fe(III) bond strength result in a cathodic shift of $E_{1/2}$, while increases in the Fe(II) bond strength shift potentials in an opposite direction. The 720-mV stabilization of TPPFeF in comparison with $TPPFeClO_4$ thus corresponds to a factor of 10^{12} increase in TPPFeF stability over that of $[TPPFeF]^-$. Finally, it has been pointed out [46] that the proposed mechanism given in scheme V applies only in noncoordinating solvents with TBAP as supporting electrolyte. Replacement of TBAP with $TBAPF_6$ or $TBAPF_4$ yields other schemes which are dominated by F^- ion produced from the supporting electrolyte.

H. VARIABLE-TEMPERATURE ELECTROCHEMISTRY

The merits of variable-temperature electrochemistry as applied to metalloporphyrin reactions are threefold. The first, and most obvious, is the ability to change the rates of any chemical reactions coupled to the electron-transfer step. In some cases reversible electrode reactions and thermodynamic values of $E_{1/2}$ may be obtained at low temperature which are not available at room temperature. Such an example is the reduction of methylcobinamide and methylcobalamin in DMF-propanol mixtures [157]. For methylcobinamide the rate constant for CH_3 cleavage after one-electron reduction was changed from 2500 sec^{-1} at 19°C to 14 sec^{-1} at $-20°C$. Reversible cyclic voltammograms could then be obtained at the lower temperature using rapid potential sweep rates. Furthermore, use of variable temperature and measurement of k_f by standard electrochemical techniques [15] allowed for calculation of the Arrhenius preexponential factor.

Another example of how variable-temperature electrochemistry may aid in mechanism determination is given by the data for oxidation of TPPFe to yield $TPPFeN_3$ in DMF [46]. At room temperature the electrooxidation product was $TPPFeN_3$, while at low temperature N_3^- was replaced by ClO_4^- from the solvent and $TPPFeClO_4$ was produced. This example has already been discussed in some detail, and current-voltage curves at several temperatures are shown in Figure 14.

The second advantage to variable-temperature cyclic voltammetry is in the measurement of electron-transfer entropies associated with the redox processes. These entropies for oxidation-reduction can be calculated directly from the temperature dependence of E^0 for the electron-transfer reactions as described by the Gibbs-Hemholtz equation

$$\Delta S_{et} = nF \frac{dE^0}{dT}, \qquad (29)$$

where ΔS_{et} is the electron-transfer entropy for the process Ox + $ne \rightleftarrows$ Red. For a reversible process in which the diffusion coefficients of the oxidized and reduced species are equal, the polarographic half-wave potential $E_{1/2}$ is a good estimate of the standard electrode potential E^0. Verification that $D_{ox} \simeq D_{red}$, and that the temperature coefficients of the oxidized and reduced species are equal, may be obtained by a measurement of the anodic and cathodic peak-current ratios at all temperatures. Thus, using the approximation that $dE_{1/2}/dT \simeq dE^0/dT$, ΔS_{et} may be obtained directly from the temperature variation of $E_{1/2}$.

The total entropy of electron transfer [obtained from Equation (29)] may be associated with changes in solvation of the central metal or the porphyrin ring, changes in spin state of the metal, changes in the number of bound axial ligands, or dimerization of the oxidized or reduced product. The assumptions involved in using Equation (29), and a separation of the potential dependence on temperature into various components, have been discussed in detail by Yee et al. [158] for the case of absolute-entropy measurements in aqueous media using a nonisothermal cell.

Using variable-temperature electrochemistry, entropies associated with the electron transfer have been measured for the reactions Fe(III) $\rightleftarrows$ Fe(II) and

Fe(II) ⇌ Fe(I) in several solvent systems including pyridine [43, 159]. Absolute values of ΔS for electron transfer may not be quantitatively compared for different electrode reactions in different solvents. It is clear, however, in the case of iron (and cobalt) reactions, which have been well studied in nonaqueous media, that a general correlation may be made between the sign and magnitude of the electron-transfer entropy and a change of axial ligation upon oxidation-reduction.

For a number of porphyrin reactions investigated to date it has been observed that, independent of the overall porphyrin charge, the entropy of electron transfer is 0 ± 5 eu when there is no change of axial ligation, and between 17 and 40 eu when one or two axial ligands are added or released upon electron transfer.* Therefore, using this simple criterion, the temperature dependence of half-wave potentials can be measured and a determination made as to whether a ligand *might be* added or released concomitant with electron transfer. This information would not be absolute, but would aid in other studies of these reactions. It should be stressed that the absolute values of ΔS^0 obtained by this method have little thermodynamic significance, since the individual contributions to the electron-transfer entropy cannot be isolated. However, as an initial step, plots of $E_{1/2}$ versus T might be constructed in order to obtain preliminary information on changes in mechanism or in axial-ligand coordination as a function of both the redox reaction and the changes in temperature.

Finally, since Fe(I) does not bind nitrogenous bases, a method is available to calculate ligand addition to Fe(II) porphyrins by measuring the differences in $dE_{1/2}/dT$ between reductions of the complexed and uncomplexed Fe(II) porphyrin. The scheme for the reaction is identical to that which has been given in scheme I and is reproduced below:

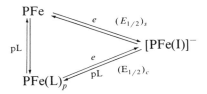

Since differences are measured between two similar electrode reactions, most errors in the liquid-junction potential of a single measurement will cancel out. The interesting and significant point is that the difference between the measured slopes $dE_{1/2}/dT$ for the two reactions directly yields the entropy of ligand addition to Fe(II). These data would be difficult to obtain under normal solution conditions. At the same time, the absolute potential difference between the two reduction processes gives a measure of the equilibrium constant

---

* Kadish, K. M., and Schaeper, D., unpublished results.

K, or free-energy change ΔG^0. When measured at several temperatures (usually between 220 and 300 K depending on the solvent), both ΔH^0 and ΔS^0 are obtained for ligand addition. Thus we have the third, and perhaps the most important, use for variable-temperature electrochemistry with iron porphyrins —calculation of the thermodynamics for ligand addition to the central metal ion.

At the time of this writing only one paper had been published which systematically investigated the thermodynamics for ligand binding of Fe(II) porphyrins using electrochemical methods [43]. In this paper values of ΔG^0, ΔH^0, and ΔS^0 are reported for ligand binding of six substituted pyridines to TPPFe(DMF) in DMF. For the reaction TPPFe(DMF) + 2L $\rightleftharpoons$ TPPFe(L)$_2$ + DMF, ΔG^0 varied from -5.6 to -8.3 kcal/mole as the pK_a of the substituted pyridine changed from 1.40 to 5.98. The enthalpy of ligand addition, ΔH^0, changed from -12.3 to -15.7 kcal/mole over the same pK_a range, while the entropy ΔS^0 was constant at 22-24 eu. For all of these ligands ΔG^0 and ΔH^0 were linearly related to pK_a, and a plot of ΔH^0 versus ΔS^0 was also linear.

I. ELECTRODE REACTIONS OF IRON DIMERS

In nonaqueous media two metalloporphyrins may be linked through a single X-bridge as shown below:

$$\text{PFe—X—FeP}$$

where X = O, N, or C. The most studied of these compounds, often by accident, is (PFe)$_2$O. This is an extremely stable complex which is formed by oxidation of Fe(II) in the presence of O$_2$ or from Fe(III) monomers in the presence of trace water or base.

i. Oxidation of Oxo-Bridged Dimers

The oxidation of (TPPFe)$_2$O and (OEPFe)$_2$O was first reported by Felton and coworkers [5, 6]. Both complexes could be reversibly oxidized in CH$_2$Cl$_2$ by single-electron-transfer steps. Redox potentials for the first oxidation step were shifted cathodically from those of the monomeric complexes by 290 mV for (TPPFe)$_2$O and 330 mV for (OEPFe)$_2$O.

Since the initial reported oxidation in CH$_2$Cl$_2$, potentials have been measured in several other solvents, but mainly those which are nonbonding. In all solvents, reversible reactions are obtained, and there appears to be little change in $E_{1/2}$ with solvent. Studies of substituent effects on $E_{1/2}$ have shown that potentials for the reversible one-electron transfers will shift in an anodic direction with electron-withdrawing substituents (as predicted) and that values of ρ calculated from Equation (23) are in agreement with values obtained for the monomeric complexes [67].

The nature of the formal oxidation state for complexes of oxidized or reduced μ-oxo dimers has not yet been resolved and is still an object of continued investigations. Based upon NMR data, magnetic measurements, and electronic absorption spectra, the oxidation product of (TPPFe)$_2$O was initially assigned as existing with one of the two iron atoms in a formal Fe(IV) oxidation state [6]. This assignment is now open to question [67, 68, 70, 160]. The latest indications appear to support radical-ion character of the product, but the data are still not conclusive.

In 1977 Wollman and Hendrickson [161] treated (TPPFe)$_2$O with HBF$_4$ in air and isolated a product which was formulated as the Fe(III)/Fe(IV) mixed-oxidation-state dimer [(TPPFe)$_2$O]$^+$ BF$_4^-$. The magnetic properties of this solid were studied as a function of temperature, and the results interpreted using a μ-oxo-bridged mixed-valence model. Later electrochemical studies by Cohen and Lavallee [162] showed conclusively that the actual product was a mixture of TPPFeF and TPPFeBF$_4$ which contained only Fe(III). In addition to the electrochemistry, earlier-reported electronic spectra and Mössbauer parameters of TPPFeF [155] agree with those for the incorrectly formulated [(TPPFe)$_2$O]$^+$ BF$_4^-$ dimer.

ii. Reduction of Oxo-Bridged Dimers

Unlike the reversible reductions in CH$_2$Cl$_2$ first reported by Wolberg [47], reduction of the μ-oxo dimers in DMF involves multielectron transfers and coupled chemical reactions associated with the electron-transfer products. Kadish, Lexa, and coworkers studied these electrochemical reactions in DMF by both electronic absorption and ESR spectroscopy [20]. In this solvent electrochemical reduction of (TPPFe)$_2$O yields first a paramagnetic ferrous-ferric dimer, then a ferrous dimer, and finally, depending on solution conditions, one of two distinct forms of Fe(I) (see earlier sections). The short-lived paramagnetic intermediate [PFe(II)—O—Fe(III)P]$^-$ was characterized at 77 K by an ESR spectrum containing a signal at $g = 1.95$. Due to the existence of overlapping reduction steps, the mixed-oxidation-state dimer could not be stabilized or isolated for further characterization in DMF. However, recent thin-layer spectroelectrochemical investigations in CH$_2$Cl$_2$ and pyridine have indicated the distinct possibility of isolating this mixed-oxidation-state intermediate, as well as a [PFe(II)—O—Fe(II)P]$^{-2}$ dimer.* Reduction in pyridine (like CH$_2$Cl$_2$) produces a well defined one electron transfer which is separated from subsequent reduction waves. Initial studies of solvent effects on (TPPFe)$_2$O redox properties indicate that there is little solvent effect on the half-wave potentials for μ-oxo dimer oxidation or reduction. Rather, the effect of changing solvents appears to be that of stabilizing or destabilizing the

---

* Rhodes, R. K., and Kadish, K. M., unpublished results.

electrochemical products of the reactions leading ultimately to dimer cleavage. The solvent independence of $E_{1/2}$ is most likely linked to the nature of the solvent interaction at the axial positions of the iron. Both the oxidized and reduced dimers appear to remain five-coordinated and devoid of axially bound solvent molecules or nitrogenous bases. This is not the case for the μ-nitrido dimer.

iii. Nitrido-Bridged Dimer

The nitrogen-bridged species μ-nitrido-bis-(α, β, γ, δ-tetraphenylporphyrin iron), written as $(TPPFe)_2N$ [163], is similar to, but not isoelectronic with, $(TPPFe)_2O$. The neutral nitrido complex is isoelectronic with the cationic species $[(TPPFe)_2O]^+$, while the reduced nitrido complex $[(TPPFe)_2N]^-$ is isoelectronic with the well-characterized 18-valence-electron$(FeTPP)_2O$. The crystal structure of $(FeTPP)_2N$ shows a similarity to both high-spin and low-spin complexes [164]. The iron atom is displaced by 0.32 Å from the plane defined by the four porphinato nitrogen atoms, compared to 0.50 Å for $(FeTPP)_2O$. In addition, the two porphinato cores in $(FeTPP)_2N$ are separated by $\simeq$ 4.15 Å, so it is shorter by $\simeq$ 0.25 Å than the μ-oxo dimers.

Electrochemical measurements in CH_2Cl_2, benzonitrile, and pyridine show that $(TPPFe)_2N$ is extremely stable in these solvents, and that up to four electrons may be extracted from, or two electrons added to, the porphyrin ring without destroying the dimeric structure. The electrode reactions of $(TPPFe)_2N$ have been reinvestigated [165], and it now appears the earlier assignment [166] of a reduction at +0.15 was in error due to compound oxidation in the workup. The neutral compound $(TPPFe)_2N$ can be reversibly oxidized at +0.15 V to yield $[(TPPFe)_2N]^+$, or reduced at −1.20 V to yield $[(TPPFe)_2N]^-$. Thin-layer spectra have been obtained for these complexes,* as well as for $[(TPPFe)_2N]^{+2}$ and $[(TPPFeL)_2N]^{+1}$ where L is a substituted pyridine or imidazole.

The site of electron transfer (ring or metal) in complexes of $(TPPFe)_2N$ is unclear at this time. In fact, it is not even certain what is the formal oxidation state of the starting species. On the Mössbauer time scale (10^{-7} sec) [163] and the ESCA time scale (10^{-15} sec) [167], the two iron atoms are equivalent and thus the dimer may be considered as consisting of two Fe(III$\frac{1}{2}$) atoms. A single sharp peak for $Fe_{2p_{3/2}}$ was obtained with a binding energy of 108.5 eV and a peak width at half height of 1.6 eV. This value of the binding energy is substantially below that of 710.5 eV for the formal Fe(III) complexes of $(TPPFe)_2O$ and might suggest a lower oxidation state in the μ-nitrido dimer.

---

* Rhodes, R. K., and Kadish, K. M., unpublished results.

Based on the x-ray PES results calling for equivalent Fe centers, the following formalisms might be proposed:

$$TPPFe(III\tfrac{1}{2})-N^{-3}-Fe(III\tfrac{1}{2})TPP, \qquad (I)$$

$$TPPFe(II)-N^{0}-Fe(II)TPP, \qquad (II)$$

$$TPPFe(III)-N^{-2}-Fe(III)TPP. \qquad (III)$$

Formalism I is that proposed by Summerville and Cohen [163]. Its assignment is difficult to reconcile with the x-ray PES results, since the binding energy of the $Fe_{2p_{3/2}}$ for $Fe(III\tfrac{1}{2})$ is expected to be greater than that of Fe(III). Formalism II contradicts the Mössbauer data. The isomer shift of iron in $(TPPFe)_2N$ is 0.099 mm/sec at 300 K and is thus comparable to shifts for other Fe(III) low-spin complexes. With formalism III one would expect to see hyperfine structure in the ESR. This was not seen by Summerville and Cohen [163] at 93 K. However, a more recent ESR study [168] in a CS_2 glass at 50 K produced well-defined ^{14}N nuclear hyperfine splittings and indicates that the unpaired electron is localized in an orbital with substantial nitrido character. This would be formalism III. Further suggestions of this formalism are also given by the resonance Raman spectra of the starting material, which give resonances indicative of high-spin Fe(III) [168].

In conclusion, the electron-transfer reactions of $(TPPFe)_2N$ are well defined in a number of solvents. What the reactions actually correspond to, and the ultimate location of the added or abstracted electron, must still be elucidated. This is not the case, however, for reactions carried out in the presence of pyridine as solvent or CH_2Cl_2-pyridine mixtures.

iv. Axial-Ligand-Binding Reactions

The starting material and the reduced dimeric products will not bind axial ligands. In contrast, rapid ligand binding of pyridine (and other nitrogenous bases) is observed for complexes of $[(TPPFe)_2N]^+$. Generation of the oxidized product, either electrochemically or chemically, yielded the same $[(TPPFeL)_2N]^+$ complex, which was characterized by both electronic absorption spectra and NMR [165]. Half-wave potential shifts as a function of added ligand concentration indicated that bis-ligated adducts of the μ-nitrido dimer are formed in solution. For eleven substituted pyridines which were investigated in 1,2-dichloroethane the magnitude of the formation constants ranged from $10^{3.8}$ to $10^{8.6}$. The magnitude of β_2 increased linearly, both with σ for the respective electron-donating or -withdrawing substituent on the coordinating ligand, and with the pK_a of the selected pyridine. This is shown in Table 13, which lists potentials and stability constants for $[(TPPFe)_2N]^+$ complexes with pyridine.

Only bis ligand adducts were observed, with no evidence for single-ligand addition, even using rapid spectroscopic measurements and low ligand con-

Table 13. Half-Wave Potentials and Stability Constants for Bis Pyridine Complexes of $(TPPFe)_2N^{+a}$ in $EtCl_2$, 0.1 M TBAP

Ligands	σ	pK_a	$\log \beta_2^b$	$E_{1/2}$ (V)^c
Neat $EtCl_2$	—	—	—	+0.17
3,5-dichloropyridine	0.75	0.67	3.8 ± 0.2	−0.06
3-cyanopyridine	0.56	1.45	4.35 ± 0.15	−0.10
4-cyanopyridine	0.66	1.86	4.74 ± 0.15	−0.12
3-chloropyridine	0.37	2.81	5.8 ± 0.2	−0.17
3-bromopyridine	0.39	2.84	5.6 ± 0.1	−0.16
3-acetylpyridine	0.38	3.18	6.90 ± 0.17	−0.24
4-acetylpyridine	0.50	3.51	6.65 ± 0.25	−0.22
aniline	—	4.63	6.89 ± 0.12	−0.24
pyridine	0.00	5.29	7.62 ± 0.04	−0.28
3-picoline	−0.07	5.79	7.8 ± 0.2	−0.30
4-picoline	−0.17	5.89	8.10 ± 0.09	−0.31
3,4-lutidine	−0.24	6.46	8.6 ± 0.2	−0.34

[a] Taken from L. A. Bottomley, K. M. Kadish, R. K. Rhodes and H. Goff, Reference [165].
[b] Reaction: $[(TPPFe)_2N]^+ + 2L \rightleftharpoons [(TPPFeL)_2N]^+$.
[c] Reaction $(TPPFe)_2N + 2L \rightleftharpoons [(TPPFeL)_2N]^+ e^-$.

centration. The reason that $K_2 \gg K_1$ is unclear at this time, but may be a change of spin state or porphyrin-bridging atom distance upon addition of the second ligand. No other form of the $(TPPFe)_2N$ complex, either oxidized or reduced, was observed to bind axial ligands.

J. COMPARISONS BETWEEN IRON COMPLEXES AND OTHER METALLOPORPHYRINS

The electron-transfer reactions of literally hundreds of different types of metalloporphyrins have been extensively reported in the literature. These have included dozens of different conjugated ring structures and over 30 different central metals of oxidation state varying between M(I) and M(VI). Both mononuclear and dinuclear metal centers have been investigated. Independent of the porphyrin selected, it is found that the thermodynamic half-wave potential, the electrochemical reversibility, and the overall mechanism are in some way influenced by the solvent composition, the degree of axial ligation, and the basicity of the porphyrin ring. Many of these relationships are common to a broad range of metalloporphyrins, while others may be specific to a specific type of complex.

Often the question arises whether conclusions regarding electrochemical reactivity and porphyrin structure may be transferred from one type of transition-metal porphyrin to a different transition-metal complex undergoing similar redox reactions at the central metal and the π-ring system. More specifically, can an understanding of iron porphyrin chemistry be gained from electrochemical studies of other transition-metal complexes? To answer this question comparisons must be made between the iron-porphyrin redox reactions and those of the same porphyrin containing a different central metal.

Because of biological implications, most electrochemical studies of metalloporphyrins have involved complexes of Fe(III) and Mn(III). The electrochemistry of Co(II) and Cr(III) porphyrin complexes has also been investigated in detail, so that there is available a good body of data comparing the reactions M(III) ⇌ M(II) as a function of solvent system, axially bound ligands, and heterogeneous electron-transfer rate constants. This will be discussed in the following sections.

i. Effect of Solvent on $E_{1/2}$ for M(III) ⇌ M(II)

The effect of solvent on half-wave potentials of TPPH$_2$ [124] and TPPCo [59, 60] has been discussed in the literature. In the former complex any shifts of potential are due to nonspecific solvent-porphyrin interactions, while in the latter complex shifts are due to differences in relative binding strength between Co(II) and Co(III). For example, CH$_2$Cl$_2$ does not bind to either Co(III) or Co(II), and the potential for the reaction Co(III) ⇌ Co(II) is observed at $+0.75$ V [60]. In contrast, both Co(III) and Co(II) bind pyridine, with stability constants for the higher-oxidation-state complex 10^{12} times greater than for the lower-oxidation-state complex [117]. This difference in stability constants leads to a substantial cathodic shift in $E_{1/2}$ from that of the uncomplexed Co(III)/Co(II). Using neat pyridine as solvent, the half-wave potential has been measured as -0.214 V versus SCE [59]. In an early comparison of the Co and Fe porphyrin reactivity it was pointed out that changes from nonbonding to bonding solvents should preferentially stabilize Co(III) over Co(II) (leading to cathodic shifts of $E_{1/2}$), while over the same solvent range, Fe(II) should be stabilized relative to Fe(III) (leading to anodic shifts of $E_{1/2}$) [59]. This statement was based on the known behavior of various Fe and Co complexes and data in the literature for TPPFeCl and TPPCo.

When the half-wave potentials for TPPCo$^+$/TPPCo are plotted against the Gutmann donor number, a linear relationship is observed for all solvents except pyridine [46, 169]. The magnitude and direction of the potential shift with solvent indicate a stronger axial coordination of solvent by Co(III). Thus, the statement regarding relative stability appears to be correct for cobalt complexes. However, since half-wave potentials for reduction of both TPPFeClO$_4$ and TPPCoClO$_4$ shift in a parallel manner with increased solvent donor number, a more correct statement is that the Fe(III) oxidation state *may be* stabilized over the Fe(II) oxidation state by solvent complexation, depending on the Fe(III) counterion.

Figure 17 shows a plot of $E_{1/2}$ versus donor number for reduction of TPPFeClO$_4$, TPPCoClO$_4$, and TPPMnClO$_4$ in CH$_2$Cl$_2$. For all three metal-centered reductions the potentials shift in a negative direction with increasing

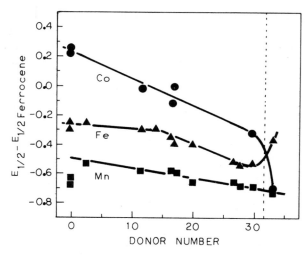

Figure 17 Plot of $E_{1/2}$ versus DN for the reduction of TPPFeClO$_4$, TPPCoClO$_4$, and TPPMnClO$_4$ in CH$_2$Cl$_2$.

donor number. This is consistent with a stabilization of the higher oxidation state over the lower oxidation state and agrees with data obtained for numerous simple inorganic systems [61].

It is interesting to note that, unlike reactions involving iron and cobalt, those for manganese are only slightly dependent on solvent [169]. This is in agreement with earlier data, taken from several laboratories, which indicate that the reactions of TPPMnX are virtually independent of solvent and only moderately dependent on counterion. The half-wave potential for TPPMnCl reduction has been reported as -0.23 V in acetonitrile [170], DMSO [171], and py [172], and -0.30 V in CH$_2$Cl$_2$ [173]. TPPMnClO$_4$ has a reduction potential of -0.19 V in EtCl$_2$ [169]. These values were not corrected for the liquid-junction potential. The difference in $E_{1/2}$ observed in Figure 17 for the reaction Mn(III) $\rightleftharpoons$ Mn(II) is less than 200 mV after correction for the liquid-junction potential. Although a case can be made for no change in $E_{1/2}$ with solvent, least-squares analysis of the data yields a negative slope. Neither CH$_2$Cl$_2$ nor EtCl$_2$ fits the relationship.

Based on the data presented in Figure 17, it appears that for solutions containing only perchlorate counterions the sensitivity of $E_{1/2}$ to solvent binding varies in the following order:

$$\text{TPPMn} < \text{TPPFe} < \text{TPPCo}.$$

It is interesting to note that pyridine does not fit the $E_{1/2}$-DN relationship for Co and Fe complexes, but does for those of Mn. No reason has been proposed for this deviation, although for each complex, substantial changes in coordination number, metal–porphyrin-plane distance, or spin state occur on going from the DMSO (DN = 28) to pyridine (DN = 33). Because py deviates in an opposite direction for complexes of TPPFeClO$_4$ and TPPFeCl (see earlier discussion), the possibility of deviations due only to uncorrected liquid-junction potential may be ruled out.

ii. Effects of Ligand Complexation on $E_{1/2}$ for M(III) ⇌ M(II)

The effect of axial-ligand coordination on half-wave potentials should be similar to those for solvent complexation, as has been discussed above. In order to compare further the reactions of TPPFeClO$_4$ with other transition-metal complexes, Kadish et al. [169] measured half-wave potentials for tetraphenylporphyrin complexes of Fe(III), Mn(III), and Cr(III) in CH$_2$Cl$_2$ containing a series of substituted pyridines. Values of $E_{1/2}$ obtained at 1.0 M ligand were then plotted versus the pK$_a$ of the free ligand. Some data are illustrated in Figure 18.

As seen from this figure, the largest shift in $E_{1/2}$ is observed for the electrode reactions of Fe, while the smallest is for the reactions involving Mn. Product stoichiometries have been determined for each of the complexes presented in Figure 18. Both Fe(III) and Fe(II), as well as Cr(III) and Cr(II), form six-coordinated bis-pyridine complexes in CH$_2$Cl$_2$ solutions containing 1.0 M ligand. In contrast, the reactants and products of the Mn(III) and Mn(II) complexes are probably five-coordinated mono-ligated species [173].

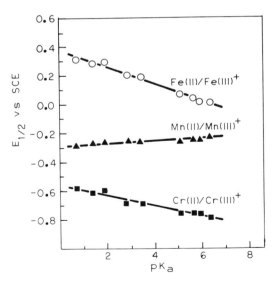

Figure 18 Plots of $E_{1/2}$ versus ligand pK$_a$ for the M(III) ⇌ M(II) reactions of TPPFe(L)$_2^+$, TPPMn(L)$^+$, and TPPCr(L)$_2^+$.

The electrode reactions of TPPFe(L)$_2^+$ have already been extensively discussed in this chapter. In CH$_2$Cl$_2$ containing substituted pyridines of pK_a > 5.4, an identical Fe(III)/Fe(II) reduction potential is observed independently of the starting material, in this case either TPPFeClO$_4$ or TPPFeCl. For these complexes, the higher Fe(III) oxidation state is more stabilized by σ-bonding than the lower Fe(II) state, and this results in a negative shift of potential with increasing pK_a of the ligand. Similarly, for oxidation of TPPCr(L)$_2^+$, the higher oxidation state is stabilized to a larger degree by σ-bonding than the lower state. There is almost no effect of ligand pK_a on $E_{1/2}$ for reduction of TPPMnClO$_4$. This independence of axial ligand implies an equal σ-stabilization of Mn(III) and Mn(II). This has been suggested in an earlier study for reactions of TPPMn(L)$^+$ Cl$^-$ where L is a substituted pyridine [173], and agrees with shifts of half-wave potential due to solvent complexation.

Similar $E_{1/2}$-pK_a plots cannot be made for the reactions of TPPCo(L)$_2^+$ in CH$_2$Cl$_2$, due to the complicated nature of the reaction Co(III)/Co(II) in CH$_2$Cl$_2$ containing 1.0 M ligand. However, for the sake of comparison, the reduction peak potentials obtained by differential pulse polarography at a platinum electrode were examined as a function of pK_a [169]. The difference in reduction potentials between TPPCo(L)$_2^+$, where L was 3,5-dichloropyridine and L was 4-N,N-dimethylaminopyridine, is 659 mV. Plots of E_{pc} vs pK_a were linear and gave a slope of $-(87$ mV$)/$pK_a. Based on this plot, the order of sensitivity to σ-effects on the oxidized and reduced forms of the investigated complexes may be listed as TPPCo > TPPFe > TPPCr > TPPMn. It should be stressed that this order refers only to the specific M(III)/M(II) couple investigated, and that substantially different effects may be observed for different electrode reactions of the same metal.

iii. Effects of Counterion on $E_{1/2}$ for M(III) ⇌ M(II)

A number of studies have shown that counterion effects on $E_{1/2}$ are substantial for the reactions of Fe(III) ⇌ Fe(II). Less data are available for the other complexes. However, sufficient data are available to have some comparison between iron porphyrins and other transition-metal complexes. M(III) ions of Mn [173] and Cr [174], like those of Fe(III), are stabilized relative to M(II) by halide complexation, and this results in a cathodic shift of $E_{1/2}$ with increasing metal-halide interaction. The largest shift is for the reduction TPPFeX, and the smallest is for TPPCrX and TPPMnX porphyrins. For both complexes there is evidence of halide ions binding to M(II), so that only relative values of M(III) stabilization may be obtained.

Similar shifts in $E_{1/2}$ due to counterion stabilization are also implied by data for the oxidation of TPPCo, where $E_{1/2} = 0.29$ V in a supporting electrolyte containing Br$^-$ and 0.94 V in a supporting electrolyte containing BF$_4^-$ [145]. The latter should bind extremely weakly to Co(III), and not at all to Co(II).

Table 14. Heterogenous Electron-Transfer Rate Constants for M(III) ⇌ M(II): Reduction of Several Complexes in DMSO

Starting Compound	k^0 (cm/sec)	Reference
OEPCrOH	2×10^{-2}	[48]
OEPMnOH	6×10^{-3}	[48]
TPPMnCl	2×10^{-3}	[171]
TPPFeCl	2×10^{-2}	[48]
TPPCo	5×10^{-3}	[48]

iv. Heterogeneous Electron-Transfer Rate Constants

Less data are available on measurements of k^0 as a function of central metal, and comparisons are more difficult between similar transition-metal complexes. Table 14 presents some data taken in DMSO for the reactions M(III) ⇌ M(II). Based on these data it is clear that there is no evident correlation between the rates of electron transfer in DMSO and the central metal ion. Correlations of k^0 with either axial-ligand binding or porphyrin-ring basicity are not as available for the other complexes as for those of the iron porphyrins.

K. CHARACTERIZATION OF REACTANTS AND PRODUCTS

In order to analyze electrode mechanisms properly it is desirable to have the electronic absorption spectra and/or ESR spectra under the same experimental conditions as the electrochemical measurement. In the former case, thin-layer spectroelectrochemistry coupled with routine electrochemical measurements provides a powerful tool in analysis. Because the path length of the thin-layer cell may be of the order of 0.1 to 0.01 cm, it is possible to directly observe the electronic spectra of metalloporphyrins at concentrations of 10^{-3} M (the ideal polarographic concentration).

Most early spectroelectrochemical measurements of porphyrins have been taken with the cytochromes [175], and there exist only a few examples in the literature of porphyrin spectroelectrochemistry in nonaqueous media. Recently the laboratory of Kadish [176] have obtained a number of thin-layer spectra for Fe(II) and Fe(I) complexes obtained as a function of potential and time of electrolysis. Many of the spectra agree with literature values obtained by conventional techniques, although in some cases transient intermediates have been detected.

An example of how the measurements may be obtained is given for the reduction of TPPFeCl in DMSO. In this solvent the Fe(III) complex exists as TPPFe(DMSO)$_2{}^+$, and the thin-layer cyclic shown in Figure 19 is obtained. In theory, the difference between the anodic and cathodic peaks should be zero

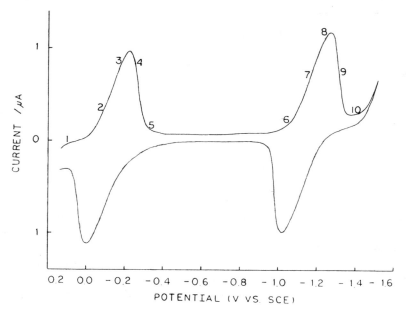

Figure 19 Thin-layer cyclic voltammogram for TPPFeClO$_4$ in DMSO, 0.1 M TBAP. Points correspond to potentials where spectra were obtained (see Figure 20).

[28]. The observed large peak separation in the cyclic voltammogram is due to uncompensated solution resistance, a problem inherent in all thin-layer cell designs [28]. In this case, supporting-electrolyte concentrations of 0.2 M are sufficient to keep the peak separation at a value not greater than 250 mV. For lower-dielectric-constant solvents, supporting-electrolyte concentrations of 0.5 M may be utilized.

Spectra were taken at points 1 to 10 on the cyclic voltammogram and are shown in Figure 20. As seen from these spectra, clean isosbestic points are obtained passing from TPPFe(DMSO)$_2^+$ to TPPFe(DMSO)$_2$ (points 1 to 5) as well as from TPPFe(DMSO)$_2$ to [TPPFe]$^-$ (points 6 to 10). Identical spectra were obtained upon multiple scans, and the starting material could be quantitatively regenerated by returning to an initial potential of -0.15 V. As already mentioned in the introduction, the main advantage of the spectroelectrochemical technique is that quantitative generation of a new species may be obtained in the absence of oxygen in a matter of seconds without the necessity of chemical oxidants or reductants. Several complete experiments may be rapidly undertaken as a function of potential without destroying the complex, and the reversibility of the system easily determined. This has not been possible in the past.

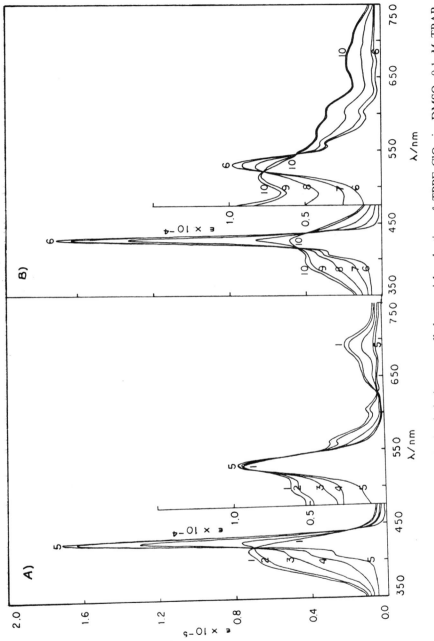

Figure 20 Thin-layer spectra obtained during controlled-potential reduction of TPPFeClO$_4$ in DMSO, 0.1 M TBAP. Numbers correspond to potentials given in Figure 19. (a) Fe(III) ⇌ Fe(II); (b) Fe(II) ⇌ Fe(I).

Acknowledgments

In this review the author has attempted to summarize the electrochemistry of Fe porphyrins in nonaqueous media. Much of the data presented have come from the author's own laboratory, since these data have often been the most extensive. It should be clear that there are now, and there have been in the past, a number of other laboratories working in the area of nonaqueous iron porphyrin electrochemistry. Their contributions have been invaluable in terms of furthering the investigations of the author's own laboratory and it is hoped they have not been slighted in this review. Finally, the author wishes to thank The National Institute of Health (Grant GM 25172) for support of much of this work.

References

1. Falk, J. E. *Porphyrins and Metalloporphyrins*. Elsevier, Amsterdam, 1964.
2. Clark, W. H. *Oxidation-Reduction Potentials of Organic Systems*. Williams and Wilkons, Baltimore, Md., 1960.
3. Stanienda, A., *Naturwissenschaften* **52** (1965) 105.
4. Stanienda, A., and Biebl, G. *Z. Phys. Chem.* (Frankfurt am Main), **52** (1967) 254.
5. Felton, R. H., Owen, G. S., Dophin, D., and Fajer, J., *J. Am. Chem. Soc.* **93** (1971) 6332.
6. Felton, R. H., Owen, G. S., Dolphin, D., Forman, A., Borg D. C., and Fajer, J. *Ann. N.Y. Acad. Sci.* **206** (1973) 504.
7. Felton, R. H., and Linschitz, H. *J. Am. Chem. Soc.* **88** (1966) 113.
8. Clack, D. W., and Hush, N. S. *J. Am. Chem. Soc.* **87** (1965) 4238.
9. Lexa, D., Momenteau, M., Mispelter, J., and Lhoste, J. M. *Bioelectrochem. Bioenerg.* **1** (1974) 108.
10. Lexa, D., Momenteau, M., and Mispelter, J. *Biochem. Biophys. Acta* **338** (1974) 151.
11. Kolthoff, I. M., and Lingane, J. J. *Polarography*, 2nd ed., 2 vols. Wiley (Interscience), New York, 1952.
12. Meites, L. *Polarographic Techniques*, 2nd ed. Wiley (Interscience), New York, 1966.
13. Sawyer, D. T., and Roberts, J. L., Jr. *Experimental Electrochemistry for Chemists*. John Wiley and Sons, New York, 1974.
14. Giraudeau, Alain. Thesis, Docteur ès Science, Physiques, Université Louis Pasteur de Strasbourg, 1978.
15. Nicholson, R. S., and Shain, I. *Anal. Chem.* **36** (1964) 706.
16. Weissberger, A., and Rossiter, B. W. *Physical Methods of Chemistry*, Part II A, Electrochemical Methods. Wiley-Interscience, New York, 1971.
17. Nicholson, R. S. *Anal. Chem.*, **37** (1965) 1351.
18. Marcus, R. A. *Electrochim. Acta*, **13** (1968) 995 and references cited therein.
19. Kadish, K. M., and Jordan, J. *Anal. Letters* **3** (1970) 113.
20. Kadish, K. M., Larson, G., Lexa, D., and Momenteau, M. *J. Am. Chem. Soc.* **97** (1975) 282.
21. Biennial Reviews, *Anal. Chem.*, 1952 through 1974.
22. Rechnitz, G. A. *Controlled Potential Analysis*. Pergamon, New York, 1963.

23. Felton, R. H. In *The Porphyrins*, D. Dolphin (ed.). Academic Press, New York, 1978.
24. Mann, C. K., and Barnes, R. K. *Electrochemical Reactions in Nonaqueous Systems*. Marcel Dekker, New York, 1970.
25. Mann, C. K. Nonaqueous Solvents for Electrochemical Use. In *Electroanalytical Chemistry*, Vol. 3, A. J. Bard (ed.). Marcel Dekker, New York, 1969.
26. Fajer, J., Borg, D. C., Forman, A., Dolphin, D., and Felton, R. H. *J. Am. Chem. Soc.* **92** (1970) 3451.
27. Murray, R. W., Heineman, W. R., and O'Dom, G. W. *Anal. Chem.* **39** (1967) 1966.
28. Hubbard, A. T., and Anson, F. C. In *Electroanalytical Chemistry*, Vol. 4, A. J. Bard (ed.). Marcel Dekker, New York, 1970.
29. Wilson, G. S., and Neri, B. P. *Ann. N.Y. Acad. Sci.* **206** (1973) 568.
30. Heiling, P., and Wilson, G. S. *Anal. Chem.* **43** (1971) 545.
31. Heiling, P., and Wilson, G. S., *Anal. Chem.* **43** (1971) 550.
32. Kadish, K. M. and Rhodes, R. K., *Inorg. Chem.* **20** (1981) 2961.
33. Ryan, M. D., and Wilson, G. S. *Anal. Chem.* **47** (1975) 885.
34. Heineman, W. R., Burnett, J. N., Murray, R. W. *Anal. Chem.* **40** (1968) 1970.
35. Kreishman, G. P., Anderson, C. W., Su, C. H., Halsall, H. B., and Heineman, W. R. *Bioelectrochem. Bioenerg.* **5** (1978) 196.
36. Hawkridge, F. M., and Ke, B. *Anal. Biochem.* **78** (1977) 76.
37. Meyer, M. L., DeAngelis, T. P., and Heineman, W. R. *Anal. Chem.* **49** (1977) 602.
38. Yildiz, A., Kissinger, P. T., and Reilley, C. N., *Anal. Chem.* **40** (1968) 1018.
39. Norvell, V. E., and Mamantov, G. *Anal. Chem.* **49** (1977) 1970.
40. Norris, B. J., Meckstroth, M. L., and Heineman, W. R. *Anal. Chem.* **48** (1976) 630.
41. Tom, G. M., and Hubbard, A. T. *Anal. Chem.* **43** (1971) 671.
42. Rhodes, R. K., and Kadish, K. M. *Anal. Chem.* **53** (1981) 1559.
43. Kadish, K. M., and Schaeper, D. *J. Chem. Soc. Chem. Comm.*, 1980, 1273.
44. Fuhrhop, J. H. In *The Porphyrins*, K. M. Smith (ed.). Elsevier, New York, 1975.
45. House, H. O., Feng, E., and Peet, N. P. *J. Org. Chem.* **36** (1971) 2371.
46. Bottomley, L. A., and Kadish, K. M. *Inorg. Chem.* **20** (1981) 1348.
47. Wolberg, A. *Isr. J. Chem.* **12** (1974) 1031.
48. Kadish, K. M., and Davis, D. G. *Ann. N.Y. Acad. Sci.* **206** (1973) 495.
49. Constant, L. A., and Davis, D. G. *Anal. Chem.* **13** (1975) 2253.
50. Kadish, K. M., Morrison, M. M., Constant, L. A., Dickens, L., and Davis, D. G. *J. Am. Chem. Soc.* **98** (1976) 8387.
51. Bottomley, L. A., Rhodes, R. K., and Kadish, K. M. In *Electron Transport and Oxygen Utilization*, C. Ho (ed.). Elsevier, New York, 1982, p. 117.
52. Kadish, K. M., and Bottomley, L. A. *Inorg. Chem.* **19** (1980) 832.
53. Headridge, J. B. *Electrochemical Techniques for Inorganic Chemists*. Academic Press, New York, 1969.
54. Vlcek, A. A. *Coll. Czech. Chem. Comm.* **16** (1951) 230.
55. Koepp, H. M., Wendt, W., and Strehlen, H. *Z. Electrochimie* **64** (1960) 483.
56. Rusina, A., Gritzner, G., and Vlcek, A. A. In *Proc. IVth Internat. Congress on Polarography*, Prague, 1966, p. 79.

57. Gritzner, G., Gutmann, V., and Schmid, R. *Electrochim. Acta* **13**, (1968) 919.
58. Bauer, D., and Beck, J. P., *Bull. Soc. Chim. France* 1973, p. 1252.
59. Walker, F. A., Beroiz, D., and Kadish, K. M. *J. Am. Chem. Soc.* **98** (1976) 3484.
60. Truxillo, L. A., and Davis, D. G. *Anal. Chem.* **47** (1975) 2260.
61. Gutmann, V. *The Donor-Acceptor Approach to Molecular Interactions*, Plenum Press, New York, 1978.
62. Brault, D., and Rougee, M. *Biochemistry* **13** (1974) 4591.
63. Kadish, K. M., and Bottomley, L. A. *J. Am. Chem. Soc.* **99** (1977) 2380.
64. Cohen, I. A., Ostfeld, D., and Lichtenstein, B. *J. Am. Chem. Soc.* **94** (1972) 4522.
65. Fuhrhop, J. H., Kadish, K. M., and Davis, D. G. *J. Am. Chem. Soc.* **95** (1973) 5140.
66. Wolberg, A., and Manassen, J. *J. Am. Chem. Soc.* **92** (1970) 2982.
67. Phillipi, M. A., Shimomura, E. T., and Goff, H. M. *Inorg. Chem.* **20** (1981) 1322.
68. Goff, H. M., Phillipi, M. A., Boersma, A. D., and Hansen, A. P. In *Electrochemical and Spectroscopic Studies on Biological Redox Components*, K. M. Kadish (ed.). ACS Advances in Chemistry Series, **201** (1982) 357.
69. Gans, P., Marchon, J. C., and Reed, C. *Nouv. J. Chemie* **5** (1981) 201.
70. Reed, C. A. In *Electrochemical and Spectroscopic Studies of Biological Redox Components*, K. M. Kadish (ed.). ACS Advances in Chemistry Series, in press, 1982.
71. Crow, D. R. *Polarography of Metal Complexes*. Academic Press, London, 1969.
72. Olson, L. W., Schaeper, D., Lançon, D., and Kadish, K. M. *J. Am. Chem. Soc.* **104** (1982) 2042.
73. Fajer, J., Fujita, I., Davis, M. S., Forman, A., Hanson, L. K., and Smith, K. In *Electrochemical and Spectroscopic Studies of Biologic Redox Components*, K. M. Kadish (ed.). ACS Advances in Chemistry Series, **201** (1982) 489.
74. Bottomley, L. A., and Kadish, K. M. *Anal. Chim. Acta* (1982) in press.
75. Rhodes, R. K., and Kadish, K. M. Submitted for publication.
76. Marchon, J. C., and Latour, L. C. Submitted for publication.
77. Wayland, B. B., and Olson, L. W. *J. Am. Chem. Soc.* **96** (1974) 6037.
78. Scheidt, W. R., and Picuilo, P. L. *J. Am. Chem. Soc.* **98** (1976) 1913.
79. Scheidt, W. R., and Brinegar, A. C., Ferro, E. B., and Kirner, J. F. *J. Am. Chem. Soc.* **99** (1977) 7315.
80. Stong, J. D., Burke, J. M., Daly, P., Wright, P., and Spiro, T. G. *J. Am. Chem. Soc.* **102** (1980) 5815.
81. Wayland, B. B., Mehne, L. F., and Swartz, J. *J. Am. Chem. Soc.* **100** (1979) 2379.
82. Stynes, D. V., and James, B. R. *J. Chem. Soc. Chem. Commun.* **325** (1973).
83. Stynes, D. V., and James, B. R. *J. Am. Chem. Soc.* **96** (1974) 2733.
84. Bottomley, L. A., Olson, L., and Kadish K. M. In *Electrochemical and Spectroscopic Studies of Biological Redox Components*, K. M. Kadish (ed.). ACS Advances In Chemistry Series, **201** (1982) .
85. Scheidt, W. R. *Acc. Chem. Res.* **10** (1977) 339.
86. Reed, C. A., Mashiko, T., Bently, S. P., Kastner, M. E., Scheidt, W. R., Sparatalion, K., and Lang, G. *J. Am. Chem. Soc.* **101** (1979) 2948.
87. Collman, J. P., Hoard, J. L., Kim, N., Lang, G., and Reed, C. A. *J. Am. Chem. Soc.* **97** (1975) 2676.

88. Goff, H., LaMar, G. N., and Reed, C. A. *J. Am. Chem. Soc.* **99** (1977) 3641.
89. Boyd, P. D. W., Buckingham, D. A., McMeeking, R. F., and Mitra, S. *Inorg. Chem.* **18** (1979) 3585.
90. Goff, A., and Shimomura, E. *J. Am. Chem. Soc.* **102** (1980) 31.
91. Brault, D., and Rougee, M. *Biochem. Biophys. Res. Commun.* **57** (1974) 654.
92. Walker, F. Ann, Lo, M. W., and Ree, M. T. *J. Am. Chem. Soc.* **98** (1976) 5552.
93. Constant, L., and Davis, D. G. *J. Electroanal. Chem.* **74** (1976) 85.
94. Wilson, G. S. *Bioelectrochem. Bioenerg.* **1** (1974) 172.
95. Harbury, H. A., Cronin, J. R., Fanger, M. W., Hettinger, T. P., Murphy, A., Meyer, J., and Vinogradov, Y. P. *Proc. Natl. Acad. Sci. U.S.A.* **54** (1965) 1658.
96. Mashiko, T., Marchon, J. C., Musser, D. T., Reed, C., Kastner, M. E., and Scheidt, W. R. *J. Am. Chem. Soc.* **101** (1979) 3653.
97. Kamen, M. D., and Horio, T. *Ann. Rev. Biochem.* **39** (1973) 673.
98. Salemme, R. F., Kraut, J., and Kamen, M. D. *J. Biol. Chem.* **248** (1973) 7701.
99. Buckingham, D. A., and Rauchfun, T. B. *J. Chem. Soc. Chem. Comm.* 705 (1978).
100. Warme P. K., and Hager, L. P. *Biochem.* **9** (1970) 1606.
101. Collman, J. P. *Acc. Chem. Res.* **10** (1977) 265.
102. Brown, G. M., Hopf, F. R., Meyer, T. J., and Whitten, D. J. *J. Am. Chem. Soc.* **97** (1975) 5385.
103. Buchler, J. W., Kokish, W., Smith, P., and Tonn, B. *Z. Naturforsch*, **33b** (1978) 1371.
104. Buchler, J. W., Kokisch, W., and Smith, P. D. In *Structure and Bonding*, Vol. 34, J. D. Dunitz (ed.). Springer-Verlag, New York, 1978.
105. Gurira, R., and Jordan, J. *Anal. Chem.* **53** (1981) 864.
106. Kadish, K. M., Schaeper, D., and Olson, L. Submitted for publication.
107. Adler, A. (ed.). *The Chemical and Physical Behavior of Porphyrin Compounds and Related Structures*, Annals of the New York Academy of Sciences, Vol. 206, 1974.
108. Caughey, W., Fulimoto, W. Y., and Johnson, B. P. *Biochem.* **5** (1966), 3830.
109. Caughey, W. S., Bralow, C. H., O'Keefe, D. H., and O'Toole, M. C. *Ann. N.Y. Acad. Sci.* **206** (1973) 296.
110. Giraudeau, A., Callot, H. J., Jordan, J., Ezahr, I., and Gross, M. *J. Am. Chem. Soc.* **101** (1979) 3857.
111. Anderson, O. P., Kopelove, A. B., and Lavallee, D. K. *Inorg. Chem.* **19** (1980) 2101.
112. Leffler, J. E., and Grunwald, E. *Rates and Equilibria of Organic Reactions*. Wiley, New York, 1963, pp. 172-179.
113. Meot-Ner, M., and Adler, A. D. *J. Am. Chem. Soc.* **94** (1972) 4763; **97** (1975) 5107.
114. Quimbley, D. J., and Longo, F. R. *J. Am. Chem. Soc.* **97** (1975) 5111.
115. Walker, F. A., Hui, E., and Walker, J. M. *J. Am. Chem. Soc.* **97** (1975) 2390.
116. Baker, E. W., Storm, C. B., McGrew, G. T., and Corwin, A. H. *Bioinorg. Chem.* **3** (1973) 49.
117. Kadish, K. M., Bottomley, L. A., and Beroiz, D. *Inorg. Chem.* **17** (1978) 1124.
118. Zuman, P. *Substituent Effects in Organic Polarography*. Plenum Publishing Co., New York, 1967.
119. Taft, R. W., Jr., *Steric Effects in Organic Chemistry*, M. S. Newman (ed.). Wiley, New York, 1956.

120. Hammett, C. P. *Physical Organic Chemistry*, 2nd ed. McGraw-Hill, New York, 1970.
121. Brown, H. C., and Okamota, Y. *J. Am. Chem. Soc.* **80** (1958) 4979.
122. Kadish, K. M., and Morrison, M. M. *Inorg. Chem.* **15** (1976) 980.
123. Kadish, K. M., and Morrison, M. M. *Bioinorg. Chem.* **7** (1977) 107.
124. Kadish, K. M., and Morrison, M. M. *J. Am. Chem. Soc.* **98** (1976) 3326.
125. Kadish, K. M., and Morrison, M. M. *Bioelectrochem. Bioenerg.* **3** (1977) 480.
126. Callot, H. J., Giraudeau, A., and Gross, M. *J. Chem. Soc., Perkin Trans. 2* **12** (1975) 1321.
127. Giraudeau, A., Ezahr, I., Gross, M., Callot, H. J., and Jordan, J. *Bioelectrochem. Bioenerg.* **3** (1976) 519.
128. Giraudeau, A., Callot, H. J., and Gross, M. *Inorg. Chem.* **18** (1979) 201.
129. Callot, H. J. *Tetrahedron Lett.* 7–8 (1973) 4987.
130. Callot, H. J. *Bull. Soc. Chim. Fr.* 5 (1974) 1492.
131. Rillema, D. P., Nagle, J. K., Barringer, L. F., Jr., and Meyer, T. J. *J. Am. Chem. Soc.* **103** (1981) 56.
132. Wolberg, A., and Mansassen, J. *Inorg. Chem.* **9** (1970) 2365.
133. Gordon, A. J., and Ford, R. A. *The Chemists Companion*, Wiley, New York, 1972.
134. Stanienda, A. *Z. Naturforsch. B* **23** (1968) 147.
135. Silvers, S. J., and Tulinsky, A. *J. Amer. Chem. Soc.*, **89** (1967) 3331.
136. Silvers, S. J., and Tulinsky, A. *J. Amer. Chem. Soc.* **86** (1964) 927.
137. Tezuma, M., Ohkatsu, Y., and Osa, T. *Bull. Chem. Soc. Japan* **49** (1976) 1435.
138. Felton, R. H. In *The Porphyrins*, D. Dolphin (ed.). Academic Press, New York, 1978.
139. Fuhrhop, J. H. In *Porphyrins and Metalloporphyrins*, K. Smith, (ed.). Elsevier, New York, 1975.
140. Kadish, K. M., and Larson, G. *Bioinorg. Chem.* **7** (1977) 95.
141. Yasuda, H., Suga, K., and Aoyagui, S. *J. Electroanal. Chem.* **86** (1975) 259.
142. Martin, R. F., and Davis, D. G. *Biochemistry* **7** (1968) 3906.
143. Kadish, K. M., and Jordan, J. *J. Electrochem. Soc.* **125** (1978) 1250.
144. Hoard, J. L., *Science* **174** (1971) 1295.
145. Truxillo, L. A., and Davis, D. G. *Anal. Chem.* **47** (1975) 2260.
146. Gygax, H. R., and Jordan, J. *Discuss. Faraday Soc.* **45** (1968) 227.
147. Bury, R., and Jordan, J. *Anal. Chem.* **49** (1977) 1573.
148. Castro, C. E. In *The Porphyrins*, D. Dolphin (ed.). Academic Press, New York, 1978.
149. Sutin, N. In *Bioinorganic Chemistry II*, Adv. Chem. Ser., **162** K. N. Raymond (ed.). American Chemical Society, 1977.
150. Feinberg, B. A., Gross, M., Kadish, K. M., Marano, R. S., Pace, S. J., and Jordan, J. *Bioelectrochem. Bioenerg.* **1** (1974) 73.
151. Davis, D. G., and Bynum, L. M. *Bioelectrochem. Bioenerg.* **2** (1975) 184.
152. Tsunchiro, T., Kallai, O. B., Swanson, R., and Dickerson, R. E. *J. Biol. Chem.* **248** (1973) 5234.
153. Margoliash, D. E., Ferguson-Miller, S., Tulloss, J., Kang, C. H., Feinberg, D. L., Brantigan, D. L., and Morrison, M. *Proc. Nat. Acad. Sci. U.S.A.* **70** (1973) 3245.
154. Brown, S. B., and Lantzke, I. R. *J. Biochem.* **115** (1969) 279.

155. Cohen, I. A., Summerville, D. A., and Hu, S. R. *J. Am. Chem. Soc.* **98** (1976) 5813.
156. Rhodes, R. K., and Kadish, K. M. Submitted for publication.
157. Lexa, D., and Saveant, J. M. *J. Am. Chem. Soc.* **100** (1978) 3220.
158. Yee, E. L., Cave, R. J., Guyer, K. L., Tyma, P. D., and Weaver, M. J. *J. Am. Chem. Soc.* **101** (1979) 1131.
159. Kadish, K. M., Thompson, L. K., Beroiz, D., and Bottomley, L. A. *A.C.S. Symposium Series* **38** (1977) 51.
160. Phillipi, M. A., and Goff, H. M. *J. Am. Chem. Soc.* **101** (1979) 7641.
161. Wollmann, R. G., and Hendrickson, D. N. *Inorg. Chem.* **16** (1977) 723.
162. Cohen, I. A., Lavallee, D. K., and Kopelove, A. B. *Inorg. Chem.* **19** (1980) 1100.
163. Summerville, D. A., and Cohen, I. A. *J. Am. Chem. Soc.* **98** (1976) 1747.
164. Scheidt, W. R., Summerville, D. A., and Cohen, I. A. *J. Am. Chem. Soc.* **98** (1976) 6623.
165. Kadish, K. M., Rhodes, R. K., Bottomley, L. A., and Goff, H. *Inorg. Chem.* **20** (1981) 3195.
166. Kadish, K. M., Cheng, J., Cohen, I. A., and Summerville, D. *A.C.S. Symposium Series* **38** (1977) 51.
167. Kadish, K. M., Bottomley, L. A., Brace, J. G., and Winograd, N. *J. Am. Chem. Soc.* **102** (1980) 4341.
168. Schick, G. A., and Bocian, D. F. *J. Am. Chem. Soc.* **102** (1980) 7984.
169. Kadish, K. M., Bottomley, L. A., Kelly, S., Schaeper, D., and Shiue, L. R. *Bioelectrochem. Bioenerg.* **8** (1981) 213.
170. Boucher, L. J., and Garber, H. K. *Inorg. Chem.* **9** (1970) 2644.
171. Kadish, K. M., Sweetland, M., and Cheng, J. S. *Inorg. Chem.* **17** (1978) 2795.
172. Kadish, K. M., and Kelly, S. *Inorg. Chem.* **21** (1982) in press.
173. Kadish, K. M., and Kelly, S. *Inorg. Chem.* **18** (1976) 2968.
174. Bottomley, L. A., and Kadish, K. M. Submitted for publication.
175. Heineman, W. R., Meckstroth, M. L., Norris, B. J., and Su, C. H. *Bioelectrochem. Bioenerg.* **6** (1979) 577 and references therein.
176. Rhodes, R. K., and Kadish, K. M. Submitted for publication.

Subject Index

Subject Index

Aquometmyoglobin 50, 51

Carbonyl Iron Porphyrins 129, 199, 208, 209
Charge Transfer Resonance 109
Controlled Potential Electrolysis 175
Coulometry 175
Cyclic Voltammetry 169
Cytochrome c 96–98, 101
Cytochrome P450 58, 141

Deoxyhemoglobin 109, 150
Donor Number 190

Electrochemical Cell for ESR Studies 178
Electrochemistry 162–249
Electron Spin Resonance 43–88
   Correlation diagram for Fe(III) 56
   EPR parameters Table 60
   EPR spectra 49–51, 69, 78, 81, 83
Electron Transfer Rate Constants 174, 219–221, 241
Electronic Spectra 97, 104

Ferric Porphyrins (Iron(III))
   Bis-pyridine complexes, formation 207
   Crystal field parameters 17
   Crystal field splitting 66
   d-orbital energies 53
   Electron transfer 174
   EPR, correlation diagram 56
   EPR, high spin 64–70
   EPR, low spin 52–64
   Iron(III)/iron(II) couple 186–191, 204, 210
   Magnetism 3–42
   Magnetic anisotropy 7, 11
   Magnetic exchange 20
   Nitrido dimer 234
   Oxidation of Fe(III) 193, 209
   Oxo-dimers 232–234
   Polarogram 167,
   Spin equilibrium 29
   Spin states 6
   Spin mixed 25, 30
   Voltammograms 170, 223, 224, 226, 242
   X-ray data 10, 26
   Zero-field splitting 16
Ferrous Porphyrins (Iron(II))
   Bis-pyridine complexes, formation 207
   Complexation, axial 195
   Iron(II)/iron(I) couple 191–199, 204
   Magnetic anisotropy 34
   One-electron energies 38
   Spin states 33
Formation constants 194–209

Half-Wave Potentials, Substituent Effects 211
Hammett Functions 213–217
Heme a 124
Hemin Chloride, magnetism 18
Hemin Reduction 220
Horse Heart Cytochrome c, EPR 50
Horseradish Peroxidase, Compound I, EPR 73–76
   Energy levels 75

Iron(IV) Porphyrins 165

Kinetic Parameters 173

Magnetism 3–42

Nitrosyl Iron Porphyrins 129, 142, 199, 202, 208, 209
   Bonding scheme 80
   EPR 77–84

Oxidation-Reduction Mechanisms 221–229
Oxyhemoglobin 108, 128, 129, 142

Porphyrins
   Axial-ligand stretching frequencies 129, 142–145
   Chromium 111, 237–241
   Cobalt 128, 129, 133, 189, 191, 237–241
   Core-size correlations 134

Electrochemistry 161–249,
Electron spin resonance 43–88
Half-wave potentials, Fe(III)/Fe(II) 187, 199, 207, 210, 218, 236
Half-wave potentials, Fe(II)/Fe(I) 191–199, 204, 207, 218
Magnetism 3–42
Manganese 106, 111, 113, 129, 133, 216, 217, 237–241
Nickel 115, 116, 124, 125, 129, 212
Out-of-plane deformations 102, 120
Oxidation state marker 140
π-π^* transitions 94, 107
Resonance Raman spectroscopy 89–159
Silver 216
Skeletal modes 114, 120, 135, 137, 145–147
Structural parameters 136
Vanadyl 140
Vinyl modes 121, 122
Zinc 140

References
 Chapter 1 39–42
 Chapter 2 85–88
 Chapter 3 152–159
 Chapter 4 244–249
Resonance Raman Spectroscopy 89–159
 Resonance Raman spectra 96, 98, 101, 103, 105, 108, 123, 148, 150, 151

Spectroelectrochemistry 180

Thiocarbonyl Iron Porphyrins 199, 208, 209
Time resolved Resonance Raman 147–151

Vibrational frequencies,
 Metalloporphyrins 129
 Axial ligand 142–145

	DATE DUE		
Chemistry Dept			

Iron porphyrins 184611